HSPA Performance and Evolution

HSPA Performance and Evolution

A Practical Perspective

Pablo Tapia, Jun Liu, Yasmin Karimli
T-Mobile USA

Martin J. Feuerstein
Polaris Wireless, USA

WILEY

A John Wiley and Sons, Ltd, Publication

Library of Congress Cataloging-in-Publication Data

HSPA performance and evolution : a practical perspective / by Pablo Tapia ... [et al.].
     p. cm.
  Includes bibliographical references and index.
  ISBN 978-0-470-69942-3 (cloth)
 1. Packet switching (Data transmission) 2. Network performance (Telecomunication) 3. Radio–Packet
transmission. I. Tapia, Pablo. II. Title: High speed packet access performance and evolution.
  TK5105.3.H73 2009
  621.382'16–dc22

                                                                                          2008052332

A catalogue record for this book is available from the British Library.

ISBN 978-0-470-69942-3 (H/B)

Typeset in 10/13pt Times by Thomson Digital, Noida, India.
Printed in Great Britain by CPI Antony Rowe, Chippenham, Wiltshire

Contents

Figures and Tables

Tables

About the Authors

 **Pablo Tapia** Pablo is a Principal Engineer in the Network Strategy team of T-Mobile USA, where he has worked in several projects including new technology evaluation, support to regulatory and business teams and technology strategy planning. He has over nine years of experience in the wireless industry, mostly focused on RAN technology efficiency and application performance. He began his career in Nokia Networks R&D, developing advanced features for GSM/EDGE networks. He has also worked as a project manager, software product manager and telecom consultant before joining T-Mobile. He holds several patents and has several academic publications, including contributions to another two books. Pablo earned a Master's degree in Telecommunications Engineering from University of Malaga (Spain).

 **Jun Liu** Jun is currently a Principal Engineer in T-Mobile USA's Network Strategy and Product Architecture group. He was the lead engineer in building the first UMTS technology evaluation network for T-Mobile USA in 2005. He has more than 10 years of experience in wireless product development and network deployment. Before joining T-Mobile USA, Jun has worked for Metawave and Western Wireless. He has two patents and many industry publications. Jun earned a BS degree in Physics from University of Science and Technology of China, a Masters and PhD degree in EE from University of Washington.

Yasmin Karimli Yasmin is currently Head of RAN Evolution and Strategy Team at T-Mobile USA. Yasmin has 15 years experience in the Telecommunications Industry starting at USWEST New Vector which then became AirTouch/Verizon Wireless and has been with T-Mobile since 2001. Yasmin led a cross functional team to evaluate and select vendors for UMTS Radio Access Network infrastructure. She and her team produced critical evaluations of T-Mobile's spectrum needs in preparation for Auction 58 (Nextwave's former 1900 MHz Assets) and the AWS (1700/2100 MHz) Auction. Yasmin has a Bachelors and Masters Degree in EE from University of Washington in Electromagnetics and Digital Communications.

Martin J. Feuerstein Marty is currently the CTO for Polaris Wireless where he leads research into wireless location technologies. He has more than 20 years of experience in research, development and deployment of wireless networks, working for companies including Lucent/AT&T Bell Labs, Nortel, Metawave, and USWEST/AirTouch/Verizon. He has consulted extensively in the industry, with many publications and more than fifteen patents in wireless telecom. Marty earned a BE degree in EE and Math from Vanderbilt University, an MS degree in EE from Northwestern University and a PhD degree in EE from Virginia Tech.

Preface

It's an exciting and fast moving time in the wireless industry with broadband access services coming to fruition. The internet and wireless are truly converging to become the twenty-first century's fundamental communications medium. Think about it. Who would have thought that customers could one day surf the Internet at DSL speeds and watch TV on their cell phones? Around the globe, a whole new generation of young people has literally grown up with cell phones and the internet. Amazingly, mobile phones have now integrated the major features of computers, cameras, camcorders, TVs, Bluetooth, WiFi and GPS devices— adding the critical elements of mobility and connectivity. These tremendous leaps have been enabled by the availability of low cost memory, high resolution displays and massive chip integration fueled in turn through the tremendous market volumes for consumer wireless devices.

Customer expectations are growing and wireless operators have to stay ahead of those expectations by offering thrilling and innovative products and services. What's the next killer application? No one can exactly predict. With new applications that could be developed for mobile devices, especially as carriers open their networks to partner developers, wireless operators must seriously improve both bandwidth and latency for data services. They are placing expensive bets on the technologies that will help them achieve these objectives. But it's not just the technology selection that is important; it's the operator's practical implementation and network optimization too. Even the best air interface technology won't be up to the immense task without the right tools and techniques to achieve coverage, capacity and quality all within the constraints of an efficient cost structure.

In this book we concentrate on extracting the most from the capabilities offered by 3GPP's HSPA radio technology, consisting of both downlink (HSDPA) and uplink (HSUPA) elements. With data rates on the downlink up to a whopping 8–10 Mbps and latencies of less that 100 milliseconds, HSPA promises to deliver the full wired internet experience to the wireless world. The big data pipe comes courtesy of extremely short time slots, fast channel quality feedback and speedy retransmissions. HSPA enables dramatically faster download times and snappier connections compared to its predecessors EDGE and GPRS, called (E)GPRS, which is great for all applications but especially demanding services like video apps. Ironically in the longer term, the real benefit may lie in the voice domain—namely high-capacity and low-latency wireless Voice over IP (VoIP) services. With technical tricks such as header compression and data DTX voice could be another data offering while increasing the overall network capacity compared to today's circuit-switched networks.

The aim of this book is to share practical implementation methods and tradeoffs for deploying, optimizing and maintaining networks using the HSPA air interface. The imperative word is 'practical', as opposed to standards, research and theory. That means we focus on real-world performance in operator's networks. We will not dive too deeply into simulation results, and we will not present theoretical derivations that you might read in research papers or in many other books written by research and development teams. Instead we will focus on lessons learned from, and techniques for optimally deploying HSPA in the field from an operator's viewpoint. We identify areas where standards have left items open for interpretation, which causes significant differences between vendor implementations. We will do so without divulging vendor proprietary algorithms, but in a way that explains what operators can expect. We also explain the essential distinctions between rolling out HSPA compared to earlier UMTS and GSM technology, because there are many issues that must be handled differently.

Our goal with this book is to help network planning and optimization engineers and managers, who work on real live networks. This is done first by setting the right performance expectations for the technology and second by sharing solutions to common problems that crop up during deployment and optimization. The book also serves as a reference for higher level managers and consultants who want to understand the real performance of the technology along with its limitations.

Foreword

It is really interesting to me that so much money and effort are being thrown around fourth-generation technologies when today's third-generation broadband wireless networks using UMTS/HSPA and CDMA2000 1xEV-DO Rev A achieve multi-megabit speeds and both are quickly being enhanced (e.g., HSPA+) to increase their throughput and capacity. This industry amazes me sometimes – people standing in line for the latest iPhone, people wanting to build out a nationwide network with free Internet access, and now the rush to 4G before 3G's potential has been fully realized.

Look at WiMAX for example. As a technology, WiMAX is on a par with HSPA and EV-DO and not light years ahead. In a megahertz-by-megahertz comparison, by everybody's measure, WiMAX has just about the same capabilities as UMTS/HSPA and EV-DO in terms of data speeds and other wireless characteristics. In the United States, I would say that today's 3G networks, the UMTS/HSPA and EV-DO networks already built by Verizon, AT&T, Alltel, Sprint, and being built by T-Mobile, are the current competitors to WiMAX mobile.

In the future, LTE will be built in phases over time. 2G and 3G systems will remain viable for many years to come and LTE will first be installed where network operators might need it in metro and industrial areas to augment their 3G data capacity. My bet is that LTE networks won't be working at full steam until 2015 or 2016. In the meantime, LTE networks will be built out in pieces on an as-needed basis. You don't build out a nationwide network in a few months, it takes a long time. This is a point I think many are missing.

If incumbent network operators are building out LTE, their customers will be equipped with multimode devices. On the 3GPP side, the devices will include GSM/UMTS/HSPA and LTE, and on the 3GPP2 side, they will include CDMA2000 1X, EV-DO, and LTE and, in some cases, all of the above. There are, and will be for a long time to come, multiple wireless standards in this world. As I have been saying for years now, customers don't care what the technology is as long as it works. And any compatibility issues will be solved within the device – with multiple slices of spectrum and multiple technologies.

All this points to the importance of network operators making maximal use of the tools at their disposal today to deliver broadband wireless – namely 3G networks using HSPA or EV-DO – and evolving those networks (as in HSPA+) to enhance customers' experiences, compete with WiMAX, and build the bridge to 4G.

Andrew M. Seybold
CEO and Principal Analyst
Andrew Seybold, Inc.

Acknowledgements

We are fortunate to work in an industry that moves at blazing speeds. It definitely adds excitement to our lives. Several of the authors are privileged to be working in the Technology Development team with T-Mobile USA at the forefront of wireless technology. We are grateful to T-Mobile USA (and Optimi where Pablo was hired from) for giving us the opportunity to learn about the topics discussed in this book.

We wish to thank the following colleagues who have collaborated with us on the projects that underlie many of the lessons reflected throughout the book: Dan Wellington, Peter Kwok, Nelson Ueng, Chris Joul, Changbo Wen, Alejandro Aguirre, Sireesha Panchagnula, Hongxiang Li, Payman Zolriasatin, Alexander Wang and Mahesh Makhijani. We would also like to extend our appreciation and gratitude to the technology leaders who served as references for this book: Harri Holma, Timo Halonen and Rafael Sanchez. Thanks for believing in us and for warning us that writing a book would be a tremendous amount of work. It was!

Last but certainly not least, a big thank you to our families for their understanding and support while we worked long hours into the night and on weekends writing this book. We dedicate this book to our young children who we hope are proud of us:

Lucia (4) – Pablo's daughter
Andrew (7) and Eric (3) – Jun's sons
Ryan (8), Daniel (5), Selene (1) – Yasmin's children
Alisa (9), Jason (7), Laura (3) – Marty's children

Albert Einstein in his wisdom advised, 'One should not pursue goals that are easily achieved. One must develop an instinct for what one can just barely achieve through one's greatest efforts.' This is the ultimate objective for our efforts on this book.

1

Introduction

There are fundamental shifts in philosophy and strategy taking place as the wireless industry matures and the power of the internet converges with the world of mobility. That appeal has drawn important new players into wireless with familiar names like Apple, Google, eBay/Skype, Yahoo!, Microsoft, Disney, CNN and ESPN. The success and innovation of social networking sites such as Facebook and MySpace have triggered numerous companies to transport these ideas to the mobile realm. The underpinning for most of these emerging areas is the widespread availability of broadband wireless accessprecisely the capability that High Speed Packet Access (HSPA) promises to deliver.

The wireless industry has reached a true crossroads with packet data services beginning to overtake traditional circuit-switched voice services. Broadband wireless access technologies such as HSPA can bring wired internet performance to the mobile domain. The combination of high data rates, low latencies and mobility enables a new generation of wireless applications not possible or even conceivable with prior technologies. In these emerging broadband wireless systems, voice itself is transported over the packet data interfaces. There are many intermediate steps involved as wireless networks transition from current circuit- to future packet-switched architectures, with HSPA and HSPA+ being two of the critical ones. Mobile service providers must efficiently master these technologies to take full advantage of broadband wireless capabilities.

With this convergence of the internet and wireless industries, the landscape has become dramatically more competitive. Broadband wireless performance is now a serious competitive differentiator in the marketplace. Customer expectations have also markedly risen, with a new generation of consumers expecting wireless systems to deliver mobile performance on par with their fixed-line DSL or cable modem home systems. To step up to that competitive challenge, wireless operators must deploy, optimize and maintain broadband wireless networks achieving dramatically higher data rates and lower latencies. This task involves not just selecting the right

HSPA Performance and Evolution   Pablo Tapia, Jun Liu, Yasmin Karimli and Martin J. Feuerstein
© 2009 John Wiley & Sons Ltd.

air interface, but also having the best possible techniques and tools to elicit optimal performance, while balancing the inevitable network quality, coverage and capacity tradeoffs.

In this book we concentrate on extracting the most from the capabilities offered by 3GPP's HSPA radio technology, consisting of both downlink (HSDPA) and uplink (HSUPA) elements. With data rates on the downlink up to a whopping 8–10 Mbps and latencies of less than 100 milliseconds, HSPA promises to deliver the full wired internet experience to the wireless world. The big data pipe comes courtesy of extremely short time slots, fast channel quality feedback and speedy retransmissions. HSPA enables dramatically faster download times and snappier connections compared to its predecessors EDGE and GPRS, called (E)GPRS, which is great for all applications but especially demanding services like video apps. Ironically in the longer term, the real benefit may lie in the voice domain – namely high-capacity and low-latency wireless Voice over IP (VoIP) services. With technical tricks such as header compression and fast dynamic power sharing, voice could be another data offering while increasing the overall network capacity compared to today's circuit-switched networks.

The wireless industry is currently observing a whirlwind of activity to invent the next big technology. The main standards groups, 3GPP, 3GPP2 and IEEE, are all hard at work on future technologies to get 4G to market – in the form of Long Term Evolution (LTE), and Mobile WiMAX. While many players in the industry are putting efforts into developing future technologies beyond 3G, we believe that the HSPA investment provides a strong and flexible platform for operators to offer highly competitive products for many years to come. In the next few years, Mobile WiMAX will enter the market as a wide area network for offering broadband wireless. From customers' perspectives both HSPA and Mobile WiMAX offer similarly high data rates and low latencies. The key advantages for HSPA are its technical maturity (being based on UMTS) and ubiquitous availability from many operators around the globe and in many devices. What many people do not realize is that for the foreseeable future, the competition for WiMAX is HSPA. Much later, when LTE enters the market in 2010+ it will likely compete with an evolved Mobile WiMAX, but it will take a number of years for LTE to reach mass scale in terms of footprint and devices. Without a doubt, HSPA will be the flag-bearer for broadband wireless services in the 3GPP world for many years to come.

The aim of this book is to share practical implementation methods and tradeoffs for deploying, optimizing and maintaining networks using the HSPA air interface. The imperative word is 'practical', as opposed to standards, research and theory. That means we focus on real-world performance in operator's networks. We will not dive too deeply into simulation results, and we will not present theoretical derivations that you might read in research papers or in many other books written by research and development teams. Instead we will focus on lessons learned from, and techniques for optimally deploying HSPA in the field from an operator's viewpoint. We identify areas where standards have left items open for interpretation, which causes significant differences between vendor implementations. We will do so without divulging vendor proprietary algorithms, but in a way that explains what operators can expect. We also explain the essential distinctions between rolling out HSPA compared to earlier UMTS and GSM technology, because there are many issues that must be handled differently.

1.1 Services and Applications for HSPA

Before diving into the technology of HSPA itself, it's worthwhile examining first the evolving data services, applications and the related ecosystem. As noted, the search is still on for a killer application for wireless data – the proverbial 'Holy Grail' that will make wireless data services as necessary to the user as the current voice service. Meanwhile, mobile operators around the world are experiencing steady increases in their networks' data usage and more importantly the associated revenues [1], driven by the launches of new network technologies and the availability of advanced Smartphone and Personal Digital Assistant (PDA) devices. This trend can clearly be seen from the growth in Figure 1.1, which shows trends in both total data revenues and the percentage of data revenues out of Average Revenue per User (ARPU).

More than relying on a single killer application, a wider adoption of data services is instead dependent on other factors, including the handset usability (user interface, processing power and speed), the capabilities of the network technology (e.g. throughput and latency), the price of the data plans, and the operators' openness to accepting new applications and partners. HSPA can certainly plan an important role in driving this adoption.

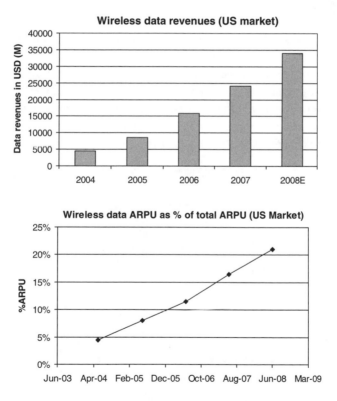

Figure 1.1 Data traffic revenue in the US 2004–2008: absolute (top) and relative to total ARPU (bottom) (data from Ref. 1)

When introducing a new air interface technology, we have found that the laggard has typically been the handset capabilities rather than the network. The reasons for this have been the severe restrictions on size, complexity, battery power, and ultimately price, which handset devices face compared to the network infrastructure. UMTS was no exception. UTRAN networks had been deployed around the world for five to six years before feature rich and stable handsets made it into the consumer hands on a mass scale.

Handsets are definitely evolving though. Today, the most basic phones have a color screen or even an embedded camera and are able to perform packet data communications. New industry players, such as Apple or Google, are creating a revolution in the market, with application rich phones that facilitate the access to applications through the wireless internet. Convergence is the word. Converged devices are starting to blossom because of the complex apps being envisioned in the future. Wireless operator's services will no longer be 'voice' or 'data', but instead a multi-media blending. Recent years have clearly demonstrated that iconic handset devices, such as the Google Phone, Apple iPhone and Motorola RAZR, can play a dramatic role in driving demand, adoption and heavy usage of new technologies. The right pairing of a handset with advanced features can indeed deliver an impressive surge in 3G penetration and data usage, as we have witnessed after launching the Google Phone in the USA. The T-Mobile G1 is an HSDPA capable phone based on the new Android operating system, creating an open mobile platform which permits users to download and install a vast array of applications created by a community ecosystem of developers across the globe. The desirable result was the generation of data traffic by the G1 that was several multiples larger than the total amount of data traffic generated by existing 3G terminals. The G1 created this impressive data dominance in an extremely short time period after its launch, handily beating out some existing handsets with embedded HSDPA capabilities, which had been launched several months before. A similar effect was experienced by the industry with the launch of the Apple iPhone, which in roughly one year has become the world's no. 2 vendor of smartphone devices, with approximately 20% of the market share (see Figure 1.2).

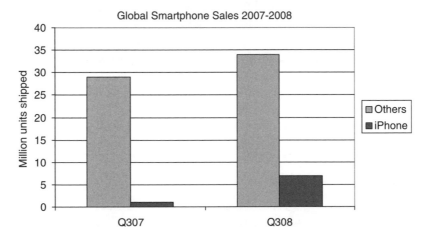

Figure 1.2 Apple iPhone sales volume since its launch in June 2007 as compared to the rest of the smartphone industry (from Ref. 2)

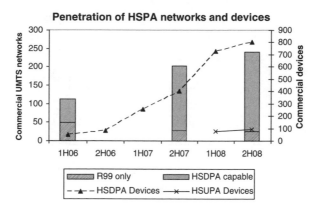

Figure 1.3 Commercial availability of HSPA 2006–2008 (from Ref. 3)

The chipset technology has also evolved and for the same price newer and more complex modems and processors are available. Following the trend of GPRS and EDGE, soon the basic chipsets for manufacturers of the 3GPP family will have HSPA modem capabilities by default. Remarkably, handsets' screen sizes are getting larger, their memory is growing and at the same time prices for those components are cheaper, which will ultimately translate into increasing demands on handsets' data transmission capabilities. As Figure 1.3 illustrates, the adoption of HSPA by both networks and handsets has been extremely rapid. Only two years after the technology was launched, 86% of the UMTS networks have HSPA capabilities with more than 800 HSDPA capable devices available in the marketplace [3,4].

With the appropriate networks and handsets ready for a true wired internet data experience, it is important to understand the characteristics of the traffic being carried because, as will be explained in more detail, this has a significant impact on how the network should be planned and dimensioned. For example, operators targeting broadband competition need to be ready to absorb a much higher demand for traffic than those with a handheld voice-centric strategy. Also, operators may decide to restrict the usage of certain services that are resource intensive, such as video streaming, in which case the capacity planning can be relaxed.

The data traffic carried by a wireless network can be further divided depending on the device used during the communication, which are typically laptops with wireless cards or handheld devices with data capabilities. In the case of computers the data traffic carried is similar to what can be found in wired broadband data networks, which is mainly based on typical internet applications like web browsing or email. Ultimately, the particular share of data traffic depends on the operator's strategy with regards to data services. For example, if the strategy is to compete in the broadband market, then more laptop-type traffic will be found, and the various parts of the networks have to be scaled accordingly. On the other hand, with a handheld-centric strategy the demand for resources will be lower in all parts of the network. In most of the cases, wireless operators carry both kinds of traffic and certain trade-offs will have to be applied, such as establishing limits to data consumption on heavy laptop users.

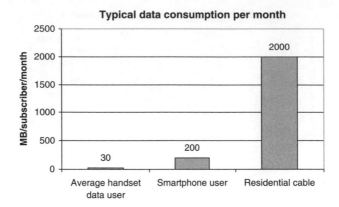

Figure 1.4 Typical data consumption depending on customer profile (type of device) compared against wired residential cable internet service

The data traffic generated by handheld devices (excluding smartphones) is typically WAP/ HTML, Blackberry email or small file transfers (e.g., MMS, ringtones, etc.), with the demand for audio and video streaming traffic remarkably increasing. The total data consumption in handheld devices is significantly lower compared to that of typical computer users, as can be seen from Figure 1.4. The figure compares average monthly data usage for wireless customers with handhelds and smartphones compared to wired internet users with residential cable service. The trend is clearly for dramatically increased wireless data usage as devices evolve and wireless broadband reaches laptops and even desktop computers.

1.2 Organization of the Book

Our goal with this book is to help network planning and optimization engineers and managers, who work on real live networks. This is done first by setting the right performance expectations for the technology and second by sharing solutions to common problems that crop up during deployment and optimization. The book also serves as a reference for higher level managers and consultants who want to understand the real performance of the technology along with its limitations.

In the second chapter we discuss the basics of UMTS and HSPA system functionality, including the standards, features, architectures and network elements. The third chapter covers typical data application performance requirements and the closely related Quality of Service (QoS) aspects of HSPA. Radio resource management, together with many of the fundamental algorithms in HSPA, are presented in Chapter 4. Chapter 5 tackles the must-know issues in the planning and optimizing of HSPA, including the key tools and techniques for rolling out networks. Radio performance of HSPA is detailed in Chapter 6, relying on numerous results from actual lab and field tests. Chapter 7 addresses management of HSPA capacity growth after initial network deployment, focussing on strategies for handling UMTS/HSPA carrier frequency expansion. The technology evolution into HSPA+ is discussed in Chapter 8, with many architecture and feature upgrades to enable better performance and longer life for HSPA. Finally, Chapter 9 presents technology strategies for UMTS/HSPA operators going forward, examining many potential evolution paths with tie-ins to LTE.

References

[1] Chetan Sharma, 'US Wireless Market Q2 2008 Update', Chetan-Sharma Technology & Strategy Consulting (http://www.chetansharma.com/), August 2008.
[2] http://www.canalys.com/pr/2008/r2008112.htm
[3] 3G Americas, HSPA status update, September 11, 2008.
[4] GSM Association, HSPA devices update, September 7, 2008.

2

Overview of UMTS/HSPA Systems

In this chapter we will provide a high level overview of the Universal Mobile Telecommunications System (UMTS) system and talk about the technology evolution and the importance of utilizing HSPA (High Speed Packet Access) in the future data network. The evolution of UMTS into HSPA has been a key factor in ensuring the competitiveness of the 3GPP technologic choice versus other alternatives such as 3GPP2 (1xEV-DO) and, more recently, WiMAX.

HSDPA represents a major improvement of the UMTS radio system for data services, with faster network dynamics and increased spectral efficiency and user throughputs. The new technology practically halved end-to-end latency values, and increased the peak rates by multiple times up to 14.4 Mbps. In the uplink side, HSUPA further improved network latency and improved the peak rates with support of up to 5.76 Mbps per user.

Section 2.1 will review the evolution of GSM (Global System for Mobile communications) into UMTS, briefly introducing the main components of a UMTS network in Section 2.2 and discussing their main functionalities. Then, Sections 2.4 and 2.5 will provide details about the new features introduced in the UMTS standards to support High Speed Packet Access (HSDPA and HSUPA).

The content of this chapter will help the readers to become familiar with the HSPA technology and its architecture before diving into more complex topics such as network performance and optimization. In later chapters, we will focus on each of the subjects touched on in this chapter and share our views on the deployment and optimization aspects from an operators' perspective.

2.1 UMTS: GSM Evolution to 3G Networks

Since its standardization in 1990 GSM has been the predominant 2G technology around the world. In 2007, the GSM Association estimated that a whopping 82% of global market was using the GSM standard with more than 2 billion users within 212 countries. The key success factor of GSM has been the open standard approach, which provided a universal technology

HSPA Performance and Evolution   Pablo Tapia, Jun Liu, Yasmin Karimli and Martin J. Feuerstein
© 2009 John Wiley & Sons Ltd.

platform for network operators across different regions around the world. Having an open standard makes the network equipment from various vendors inter-operable and provides tremendous operation benefits and cost savings for the network operators. These cost savings are in turn transferred to the consumers, providing them with the flexibility to move from one operator to the other without changing their handsets.

In the late 1990s, with the wireless industry trends slowly shifting from voice to data services, the GSM industry decided to design a new network technology that would support faster and more efficient data transmissions. This new technology, called Universal Mobile Telecommunications System (UMTS), would be based on the same open concepts as GSM and would also benefit from the economies of scale resulting from having multiple operators around the world agree on the same evolution path. The global success of GSM demanded that the 3G standard be backwards compatible to ensure a smooth technology migration for both network operators and end users.

Following the trend from GSM, UMTS is today the most extended 3G technology worldwide with 248 deployments, 223 of those including support of High Speed Data services (HSDPA or HSDPA+HSUPA).

2.1.1 Overview of UMTS Standardization

In 1998 a major standards organization called 3GPP ('The 3rd Generation Partnership Project'), was formed to develop technical specifications for the '3rd Generation' or 3G Network for the GSM technology path. 3GPP is a common forum for several standard development organizations, including the European Telecommunications Standards Institute (ETSI), the American National Standards Institute (ANSI), the Chinese Wireless Telecommunication Standard (CWTS), the Telecommunications Technology Association (TTA) and the Association of Radio Industries and Business/Telecommunication Technology Committee (ARIB/TTC). The agreement behind this collaborative effort was that all standards work on the GSM family of technologies should be done within 3GPP rather than within individual regional standards bodies as had been done in the past.

The first release of the UMTS standard was finalized in 1999, covering the Radio Access Network (RAN), Core Network (CN), Terminals, Services and System Aspects and the migration path from GSM with the incorporation of GERAN (GSM EDGE Radio Access Network). The first release of the UMTS standards is generally referred to as Release'99 or Rel.'99. Due to the importance and large scale deployment of GSM, the UMTS standard put a strong emphasis on the backward compatibility with its 2G system, as well as the flexibility to incorporate more advanced features in future developments. Another requirement for Rel.'99 UMTS was to minimize the impacts on the Core Network when introducing the UTRAN, and basically utilize many of the existing elements from the GSM network (GPRS core, HLR, VLR, etc.).

Since the introduction of UMTS Release'99, the standards have been evolving with multiple major releases:

- Rel.'5 introduced High Speed Downlink Packet Access (HSDPA), an enhancement that permits the increase of downlink data rates, capacity and significantly reduces latency.

- Rel.'6 introduced High Speed Uplink Packet Access (HSUPA), which incorporates similar improvements on the uplink direction.
- Rel.'7 and Rel.'8 provide further improvements to HSDPA and HSUPA, with a strong focus on application performance. These last two releases will be discussed in more detail in Chapter 8 (HSPA Evolution).

2.1.2 UMTS Network Architecture

The UMTS network architecture bears a lot of similarities with that of GSM as shown in Figure 2.1. The main components of the UMTS network are the UTRAN (Radio Access) and the Core Network. The Radio Access Part (UTRAN) takes care of the radio transmission and reception, including segmentation, error protection and general radio resource management, among other things.

The Core Network is further divided into Circuit Switched (CS) and Packet Switched (PS) network. The packet core network is the one involved in data communications, and is composed of two main nodes: the Serving GPRS Support Node (SGSN) and the Gateway GPRS Support Node's (GGSN). The SGSN performs PS mobility management, packet routing and transfer, authentication and charging. The GGSN's main functionality is the routing of packets to and from external networks. It also performs authentication and charging functions, and is the main node responsible for Quality of Service (QoS) negotiation.

Within the UTRAN, a Radio Network Subsystem (RNS) contains the Radio Network Controller (RNC) and its corresponding Node-Bs. The RNS is equivalent to the Base Station Subsystem (BSS) in GSM. Each RNS is hosted by the core network through the Iu interfaces (IuCS for voice and IuPS for data). Its counterparts in a GSM system are the A and G interfaces

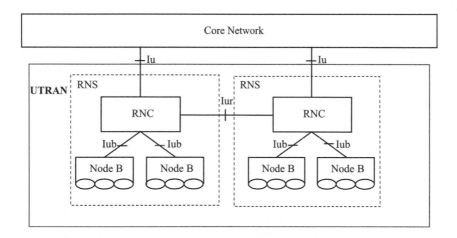

Figure 2.1   UTRAN architecture [1] © 2008 3GPP

for CS and PS respectively. Node-B and RNC are connected through the Iub interface, which is similar to the Abis interface between the BTS and BSC of a GSM system.

UTRAN defines a new interface between different RNCs for mobility control over the RNC boundaries, the Iur. The Iur interface enables soft handovers along the RNC boundaries. There is not an equivalent interface in GSM.

2.1.3 Air Interface Technology

The most important improvement introduced by UMTS is the air interface. The technology selected for the air interface is called Wideband Code Division Multiple Access (WCDMA). The following are the key characteristics of the WCDMA technology as compared to the GSM air interface:

- frequency reuse one;
- spread spectrum technology;
- use of fast power control;
- use of soft handover.

Unlike GSM, which is based on Time Domain Multiple Access (TDMA) technology, WCDMA applies spread-spectrum modulations schemes in which user data (typically with small bandwidth) is spread over a channel with large bandwidth. All users share the same spreading spectrum and each user is identified by a unique channelization code assigned by the system. The channelization codes are orthogonal to each other to ensure the independency of each channel. Figure 2.2 shows the difference between CDMA and TDMA frequency utilization scheme.

2.1.3.1 Power Control

Since all users share the same frequency in a WCDMA system, it becomes absolutely essential to manage interference from all radio links. A fast power control mechanism operating at 1500 Hz helps control the power transmitted by each of the users, making sure they use *just* the amount of power they need, and at the same time improving the link protection against fast fading.

2.1.3.2 Spreading Gain

WCDMA can operate either in FDD or TDD modes, but for the purpose of this book and sharing lessons learnt, only FDD will be discussed. The carrier bandwidth for a FDD WCDMA system is 2×5 MHz with a chip rate of 3.84 Mcps. Using a wide carrier frequency band to spread the user data reduces the frequency selectivity of the radio channel; that is, the radio channel becomes more resilient to multipath reflections, noise and interference. This typically is characterized as the spreading gain or processing gain of a CDMA system.

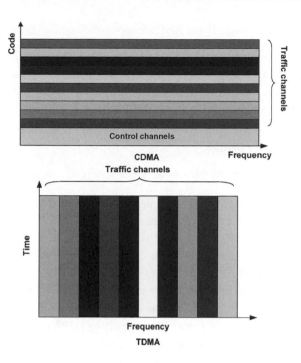

Figure 2.2 CDMA vs. TDMA: Different frequency utilization scheme

Let's consider the AMR 12.2 k voice service as an example. In this case, the user data rate is 12.2 kbps; since the chip rate is 3.84 Mcps, the spreading gain from CDMA will be $3.84/12.2^*1000{=}24.9$ dB. As we can see, for different user data rates, the spreading gain will be different. The higher the user data rate, the less the spreading gain. In terms of coverage, this means that services demanding higher data rates will have smaller coverage footprint as illustrated in Figure 2.3.

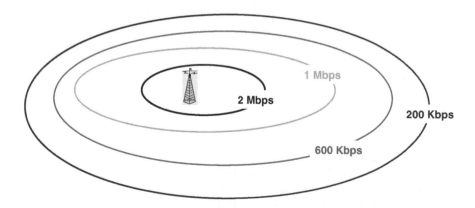

Figure 2.3 UMTS coverage for services with different data rate

2.1.3.3 Soft Handover

Soft handover is a unique feature in the CDMA technology. The capability of keeping at least one active radio link during the cell transition provides significant benefits for UE mobility performance. Soft handover also takes advantage of macro-diversity combining of multiple radio links and improves the capacity and performance of the system.

2.2 UMTS System Elements

2.2.1 User Equipment (UE)

In UMTS the term UE is the equivalent of the mobile device in GSM. Although the concept of UE is not limited to handset or data cards, in the context of this book, the main focus will be on these conventional devices, which in general will be carrying the majority of the 3G traffic.

Most UMTS UEs are multi-band dual mode devices, i.e. supporting both GSM and UMTS, to ensure backward compatibility with the 2G network. In today's devices, a very common combination is quad band GSM, plus dual band UMTS support. This allows users to travel from network to network without noticing the technology or the band changes.

In recent years, there are more and more UEs supporting HSPA capability. With the increase of data traffic demand and reduction of the chipset prices, it can be expected that most of the UMTS terminals will support HSPA in the near future.

2.2.2 Node-B

The Node-B is the equivalent of the GSM Base Station (BTS). It hosts all base band equipment (such as channel elements and processing boards), radio components and power system. In general, all these are called radio resources which play a central role in the planning and optimization process of the UMTS network. One Node-B will typically serve three sectors, and each of these can be configured with multiple carriers (frequency channels). The Node-Bs are connected to a centralized node, called Radio Network Controller (RNC) via the Iub interface.

Although all radio resources such as channel element and power are provided by the Node-B, most of the radio resource management functions in UMTS Rel.'99 reside in the RNC. However, as we will see later on in this chapter, with the introduction of HSDPA and HSUPA, many of these functions will be moved to the Node-B.

2.2.3 Radio Network Controller (RNC)

The RNC is the brain of the Radio Access Network (RAN) and is responsible for the Radio Resource Management (RRM) of the cells. The RNC performs functions such as admission control and congestion control; it also manages the UE mobility and macro diversity combining on the uplink, which enables soft handovers. With the support of Iur interface, one UE can be connected to more than one RNC at a time. In such a scenario, the RNC controlling the signaling and the link radio resources is the 'Serving RNC' (SRNC), while the new RNC is called 'Drift RNC' (DRNC) and will support the Serving RNC by providing network resources through the Iur interface. Serving RNC relocation and soft handover over Iur are the main functions for mobility control along the RNC boundaries.

2.3 UMTS Radio Bearers and Services

In this section, we will analyze the service richness that the UMTS technology offers via different combinations of radio bearers. The standardization of UMTS put a significant effort in designing new bearer capabilities for UTRAN, which are defined through a set of bearer attributes and possible values. The following main categories were defined for the radio interface related specifications:

- *Information transfer attributes.*
- *Quality of Service (QoS) attributes.*

2.3.1 Information Transfer Attributes

The information transfer attributes define the requirements for transferring user information between multiple access points. These include: information transfer rate attributes (peak bit rate, mean bit rate, occupancy), connection mode attribute, symmetry attribute and communication configuration attribute.

UTRAN is designed to ensure the efficient support of a wide variety of UMTS services, including future services which could be defined as new applications become available. Each radio bearer can be mapped to one or more logical channel to support variable data rate services in which the rate is set at the user request or as determined by the network capabilities. The following are the bit rate requirements that need to be supported under different radio conditions:

- at least 144 kbits/s in satellite radio environment;
- at least 144 kbits/s in rural outdoor radio environment;
- at least 384 kbits/s in urban/suburban outdoor radio environments;
- at least 2048 kbits/s in indoor/low range outdoor radio environment.

It is also required that terminals shall support multiple bearer services simultaneously in UTRAN.

2.3.2 Quality of Service (QoS) Attributes

The Quality of Service (QoS) attributes set the performance thresholds such as maximum bit rate error, delay variation tolerance, maximum transfer delay etc. The support of Quality of Service permits different traffic flows to be treated with the appropriate attributes (delay, jitter, bit error rate) required by the application. For example, when initiating a Voice over IP (VoIP) call, the network will set the necessary delay requirements for a real-time service (i.e., voice), which will be different from the delay requirements for a web browsing session. The benefits of QoS and its supporting functionality in UMTS will be discussed in detail in Chapter 3.

2.4 HSDPA (High Speed Downlink Packet Access)

Rel.'99 was an important milestone in the development of the 3GPP standards. It ensured a smooth 3G migration for the GSM network. The introduction of UTRAN provides substantial performance enhancements over the GSM system. However, since a lot of attention was given to the backward compatibility with the existing GSM core network, many compromises were made in order not to change the core network structure.

From a radio spectral efficiency standpoint, Rel.'99 is very primitive because it dedicates channel resources to each data user, and power control mechanisms are the same between real-time applications such as voice and non real-time data. Dedicating channel resources for each data user is inefficient. Since data is bursty in nature, implementations which can leverage the user multiplexing are more efficient at delivering data: that is the reason why HSDPA was introduced to improve the offering of data services on UMTS networks.

With the introduction of the High Speed Downlink Shared Channel (HS-DSCH), fast HSDPA scheduler, adaptive modulation and coding and fast retransmission (HARQ), Rel.'5 HSDPA is able to deliver improved downlink cell throughput up to 14.4 Mbps. In addition to the data rate improvement, HSDPA also reduces network latency. From an end user's perspective, these can be directly translated into improved user experience.

The introduction of HSDPA has been very successful because it greatly improves the downlink performance with higher throughput and lower latency than (E)GPRS and provides network operators an opportunity for incremental data revenues. The improved performance and efficiency of HSDPA permit the operators to adequately serve the growing market demand for higher data volumes and richer applications. In the next section, we will briefly go through the key aspects of HSDPA to give readers a high level view of the design philosophy of HSDPA. In later chapters, we will deep dive into each of those subjects to gain a full understanding of the technology.

2.4.1 Motivation for the Introduction of HSDPA

Before jumping into the discussion on HSDPA, it is necessary to spend some time first understanding why Rel.'99 couldn't meet future data demand and what kind of improvements were expected from HSDPA.

There are two major factors that contributed to the inefficiency of Rel.'99 data service:

1. Dedicated channel assignment for each data user; and
2. Centralized control of the resources at the RNC.

As we know, data applications have different traffic characteristics compared with voice. Circuit switched voice traffic requires dedicated channels with fixed, small bandwidths. It demands continuous transmission and stringent delay budget to meet the real-time quality of service requirement. Data traffic, on the other hand, is typically more bursty and highly dependent on the type of application running on the mobile device or PC. Therefore it is imperative that the assignment of radio resources for data users be flexible and adapt quickly to the changing radio conditions and application demands.

With Rel.'99 traffic, the packet scheduling function for data resides in the RNC. Since the RNC makes decisions based on the measurement reports sent by the Node-B (typically every 100 ms) and enough measurement reports are required to make an educated decision, the frequency at which the packet scheduler can make an adjustment could be lower than that of the data traffic fluctuations. Slower schedulers lead to lower efficiency in utilizing the system resource and poor user experience (low throughput and long latency). This is one area that HSDPA will improve over Rel.'99.

In addition, the channel structures and flow control in Rel.'99 are similar for voice and data. Each data user has its dedicated control and traffic channels. The resources assigned to a specific user, such as channel element, power and code, cannot be shared with other users even when the application is not sending or receiving data. This is also a source of inefficiencies that HSDPA will solve through the introduction of a new channel, called HS-DSCH, that is shared among all the users in the cell and that employs Time Division Multiplexing (TDM) in addition to CDMA.

2.4.2 Main HSDPA Features

The introduction of HSDPA represented a revolution of the existing WCDMA standards, with significant impacts in several parts of the Radio Access Network. The key modifications introduced were:

- introduction of a shared data channel, multiplexed in time (TDM);
- introduction of new modulation (16QAM) and coding schemes, with the possibility to transmit with up to 15 parallel channelization codes;
- modification of the MAC protocol architecture to enable faster response to changes in user demands and radio conditions;
- introduction of Adaptive Modulation and Coding (AMC) and new error correction mechanisms in the MAC layer (HARQ).

One important improvement brought by HSDPA is the latency reduction. As we will discuss in detail in further chapters (see Chapter 5), two operators' networks can have the same peak throughput, yet the one with lower end-to-end latency will provide better user experience. We have noted that many performance managers and network designers in different wireless companies use peak throughput when designing, evaluating or optimizing their data network.

In addition to the new traffic channel, HSDPA modifies the architecture of the radio protocols to better manage the resources provided by the shared channel. Adaptive modulation and coding (AMC), hybrid automatic repeat request (HARQ) and fast packet scheduling complement the offer, taking full advantage of the faster network dynamics to improve the efficiency of the data transmissions. In the following sections, we will cover the main HSDPA features, one by one.

2.4.2.1 HSDPA Channel Structure

High speed downlink packet access (HSDPA) was introduced in Rel.'5 to provide enhancements over Rel.'99 to compete with EVDO and WiMAX. To increase the channel utilization efficiency, HSDPA introduces a new physical channel called high speed downlink shared channel (HS-DSCH). This channel is shared among all data users and uses a 2 ms frame or TTI (Transmission Time Interval), as compared to the 10 ms radio frame used by Rel.'99 channels. The shared channel is a completely new concept compared with the conventional UMTS Rel.'99 in which each user had dedicated resources. A big shared data pipe provides the benefit of statistical multiplexing among all data users and thus improves the overall system efficiency. Along with this new channel, there are several associated channels defined in the standard as well. Table 2.1 lists all these channels and their functions.

2.4.2.2 New Modulation and Coding Schemes

While Rel.'99 only supported one type of modulation (QPSK), in HSDPA three different modulation schemes have been defined: QPSK, 16QAM and 64QAM. Typical devices today support QPSK and 16QAM, and at least five parallel codes. With five codes allocated to HSDPA, QPSK can provide 1.8 Mbps peak rate and 16QAM can reach 3.6 Mbps. Table 2.2 lists all UE categories which are supported by the HSDPA network.

2.4.2.3 Modified Protocol Architecture

Rel'5 introduced a modification of the radio protocol architecture. The MAC layer was split in two sublayers, each of them located in a different network element: MAC-d, located in the RNC and MAC-hs, located in the Node-B. The data to be transmitted on the HS-DSCH channel is transferred from Mac-d to Mac-hs via the Iub interface. A Mac-hs entity has four major functionality components:

- Flow control, which manages the data flow control between Mac-d and Mac-hs;
- Scheduler, which manages the HSDPA resources in the sector;

Table 2.1 New channels introduced for HSDPA

	Channel	Description
Downlink	HS-DSCH	High Speed Downlink Shared Channel. Transport channel carrying user plan data.
	HS-SCCH	High Speed Shared Control Channel. Common control channel for HSDPA. Carrying information such as modulation, UE identity etc.
Uplink	HS-DPCCH	CQI reporting, HARQ Ack/NAck

Table 2.2 HSDPA UE category defined by 3GPP

HS-DSCH category	Maximum number of HS-DSCH codes received	Minimum inter-TTI interval	Maximum number of bits of an HS-DSCH transport block received within an HS-DSCH TTI	Total number of soft channel bits	Supported modulations without MIMO operation	Supported modulations simultaneous with MIMO operation
Category 1	5	3	7298	19200	QPSK, 16QAM	Not applicable (MIMO not supported)
Category 2	5	3	7298	28800		
Category 3	5	2	7298	28800		
Category 4	5	2	7298	38400		
Category 5	5	1	7298	57600		
Category 6	5	1	7298	67200		
Category 7	10	1	14411	115200		
Category 8	10	1	14411	134400		
Category 9	15	1	20251	172800		
Category 10	15	1	27952	172800		
Category 11	5	2	3630	14400	QPSK	
Category 12	5	1	3630	28800		
Category 13	15	1	35280	259200	QPSK, 16QAM, 64QAM	
Category 14	15	1	42192	259200	QPSK, 16QAM, 64QAM	
Category 15	15	1	23370	345600		QPSK, 16QAM
Category 16	15	1	27952	345600		QPSK, 16QAM
Category 17	15	1	35280 23370	259200 345600	QPSK, 16QAM, 64QAM –	– QPSK, 16QAM
Category 18	15	1	42192 27952	259200 345600	QPSK, 16QAM, 64QAM –	– QPSK, 16QAM
Category 19	For future use; supports the capabilities of category 17 in this version of the protocol					
Category 20	For future use; supports the capabilities of category 18 in this version of the protocol					

- HARQ entity;
- Transport Format and Resource Combination selection entity.

The major benefit from the modified MAC architecture is the possibility of controlling radio resources at the Node-B level, which provides a fast and accurate response to the changing radio environment. A key element of HSDPA is the Node-B based fast packet scheduler. The HSDPA scheduler is able to look at each user's channel condition based on the channel quality reported by the UE and assigns code and power resources accordingly on each 2 ms TTI. This enables the system to allocate resource to each user quickly.

Chapter 4 will provide details on the different types of scheduling strategies. Also, Chapter 6, will discuss the performance of each of the schedulers under different radio conditions and application profiles. We will show that the efficiency of the HSDPA scheduler is tied to the application profiles and related QoS requirements.

2.4.2.4 Adaptive Modulation and Coding (AMC)

Adaptive modulation and coding (AMC) provides an alternative to the conventional CDMA fast power control for link adaptation. With AMC, the coding rate and modulation scheme for each user is dynamically adjusted based on the average channel condition of the radio link. In general, the channel power is constant over an interval defined by the system. The link adaptation is achieved by assigning a different modulation scheme or coding rate to each user based on the channel condition reported by the terminal device. The channel quality index (CQI) has been defined in the standard to report the channel condition measured by the mobile to the HSDPA scheduler. However, the calculation of the CQI value is not clearly defined in the standard, and hence variations among handset vendors in CQI reporting can be expected. Chapter 4 will provide more details on the mechanisms controlling AMC.

Note that for AMC to work effectively, timely channel condition reporting and quick packet scheduling are essential. These requirements, however, cannot be satisfied by the radio link control (RLC) level based packet scheduler defined in Rel.'99. The Mac-hs, which is a new protocol in Rel.'5 resides in the Node-B is intended to provide such enhancements. HARQ and fast packet scheduler combined with AMC are the core of this new development.

2.4.2.5 HARQ (Hybrid Automatic Repeat ReQuest)

HARQ improves throughput by combining failed transmission attempts with the re-transmissions, effectively creating a more powerful error correction scheme. Hybrid ARQ can be combined with Adaptive Modulation and Coding (AMC), making the initial selection of modulation and code rate more tolerant to errors.

HSDPA defines two different Hybrid ARQ methods: Chase Combining and Incremental Redundancy (IR), that will be selected based on the UE category and network configuration. With Chase Combining, the re-transmitted data is identical to the original transmission, whereas with Incremental Redundancy, each retransmission provides new code bits from the original code to build a lower rate code.

There are pros and cons on both algorithms. Chase Combining is sufficient to make AMC robust and consumes less buffer memory while IR offers the potential for better performance with high initial code rates at the cost of additional memory and decoding complexity in the handset.

HSDPA utilizes a HARQ protocol based on N channel 'Stop And Wait' (SAW) scheme. The utilization of multiple parallel processes allows data to be continued to be sent through the link even when one packet has not been properly received in one of the channels. The number of parallel processes, as well as the length of the radio frame, have an impact on the latency in the radio network.

Figure 2.4 below shows a typical response time for the HARQ process, which illustrates the dependency between the number of simultaneous processes (buffers) and the delay of the network.

The diagram in Figure 2.4 shows that the more users on the Node-B, the slower the response time. In the figure, T_{Prop} is the propagation time over the air, T_{UEP} is the UE processing time, T_{ACK} is the period when the acknowledgment is sent, T_{NBP} is the Node-B processing time and T_{CTRL} is the time for sending UE the control information which is related to the data sent on the shared channel.

The total processing time for each UE and the network is the time when the system is dealing with other sub channels and the UE is not receiving information or sending acknowledgment to the network.

$$T_{\text{Processing}} = T_{\text{UEP}} + T_{\text{NBP}} + T_{\text{CTRL}}$$

As we can see, both the TTI length and number of sub channels have an impact on the latency. Shorter TTI provides short latency but leaves less processing time for UE and the network. More sub channels means longer latency but provides more processing time. So there is a balance between the performance and the system capability. Table 2.3 illustrates the processing times for different TTI/sub channel combinations [2].

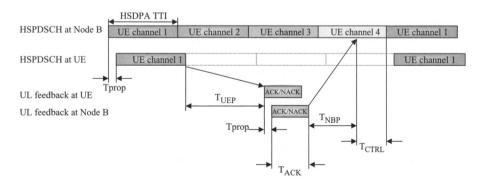

Figure 2.4   Four-Channel SAW HARQ © 2008 3GPP

Table 2.3 Processing time for UE and network for SAW HARQ

	2 sub channels				4 sub channels			
TTI length	1-slot	3-slot	5-slot	15-slot	1-slot	3-slot	5-slot	15-slot
$T_{process}$	0 ms	1.33 ms	2.67 ms	9.33 ms	1.33 ms	5.33 ms	9.33 ms	29.33 ms

© 2008 3GPP

HSDPA is based on a three-timeslot TTI structure (2 ms = 0.667*3). Typically the number of HARQ sub channels is six. The number of parallel HARQ processes that an UE can handle depends on the UE category as defined in Table 2.4. The table also lists the size of the virtual buffer of each HARQ process.

2.5 HSUPA (High Speed Uplink Packet Access)

It has been generally accepted that internet traffic has an asymmetric nature, that is, users are more likely to download content from the internet than to upload their own information. Such a trend was determined by the wide utilization of applications such as web browsing, content downloading (music, documents, etc.) and audio or video broadcasting which predominantly generate traffic on the downlink. However, with the emergence of more social networking applications such as YouTube and MySpace, the demand on uplink capacity has been steadily increasing. Users will upload information, typically multimedia, from their own PC or other consumer electronic devices to the internet on a regular basis and share them with other members of the virtual community they belong to. In today's mobile industry, mobile handsets are trending toward multi-functional consumer electronic devices. New smartphones such as the iPhone or the G-Phone represent good examples of this trend. Applications such as online gaming, video sharing and other Multi Media Services are becoming more and more popular, especially among the younger generations. The uplink demand from mobile applications has forced the industry to search for an uplink data solution which can match the performance of HSDPA.

2.5.1 Main HSUPA Features

In 3GPP release 6, enhanced uplink performance, also known as HSUPA, was proposed to improve the uplink dedicated transport channel performance for data services. The goal was to provide significant enhancements in terms of user experience with increased throughput, lower latency and higher capacity.

Table 2.4 Number of HARQ processes supported by different UE category

UE Category	1	2	3	4	5	6	7	8	9	10	11	12
# of HARQ	2	2	3	3	6	6	6	6	6	6	6	6

HSUPA enhanced uplink includes a set of new features, very similar to the ones introduced previously by HSDPA in the downlink direction. The impact of these changes on the existing protocols and network structure was not significant:

- Introduction of new coding schemes, with the possibility of transmitting up to four parallel channelization codes'.
- Modified MAC architecture to improve the system dynamics. This modification introduces a shorter TTI (of up to 2 ms), and permits the utilization of fast Node-B schedulers.
- Fast retransmission schemes with AMC and Hybrid ARQ between the UE and the Node-B.
- Higher order modulations such as 16QAM (introduced in Release 7).

Although some of the features introduced by HSUPA are similar to those in HSDPA (short TTI, Node-B based fast retransmission and fast scheduling), HSUPA is fundamentally different because of the usage of a dedicated channel per user, called E-DCH. This results in very different treatment of radio resources.

The traffic scheduling is rather random in HSUPA since all data users are not synchronized with each other over the air interface from the scheduler point of view, and therefore a common pipe cannot be established by the network for all users as it does for the downlink. Since E-DCH is a dedicated channel for each data user, the chances are that multiple UEs may be transmitting at the same time on the uplink and causing interference for each other. Therefore the Node-B scheduler must control the transmit power level of each individual user to ensure that all uplink data traffic can coexist within the system capacity limit. From this perspective, the HSUPA scheduler is fundamentally different from that of the HSDPA. The HSDPA scheduler assigns the shared pipe among multiple users based on the buffer size, channel condition and traffic priority, whereas the HSUPA scheduler primarily uses a power control scheme to manage the uplink noise rise. Chapter 4 will provide more details on the scheduling mechanisms in HSUPA.

Table 2.5 summarizes the major differences between HSDPA and HSUPA up to Rel.'6.

Table 2.5 Differences between HSDPA and HSUPA

Features	HSDPA	HSUPA
Maximum throughput	14.4 Mbps	5.76 Mbps
Node-B based scheduling	Yes	Yes
TTI	2 ms	2 or 10 ms
Modulation	16QAM and QPSK	QPSK
HARQ	Yes	Yes
Transport channel	Shared HS-DSCH	Dedicated E-DCH
Scheduling scheme	Rate control	Power control
AMC	Yes	No
Handover	Cell change	Soft handover, but serving cell follows HSDPA

Since some of the concepts previously introduced for HSDPA are also applicable to HSUPA (such as HARQ), the following sections will focus on the main changes introduced by HSUPA: the new channel structure and coding schemes, and the modified MAC architecture.

2.5.1.1 HSUPA Channel Structure and Coding Schemes

The E-DCH maps to two new physical channels on the uplink: E-DPDCH carrying the traffic information and E-DPCCH carrying the control information. The control information includes the 'happy bit' which is used by the UE to signal to the Node-B whether it is happy with the scheduled rate or not. The transport format information (e-TFCI) is also carried by E-DPCCH. There are three new channels on the downlink to ask the UE if it can power up or down during a call. E-AGCH, the absolute grant channel, sends an absolute grant from the Node-B scheduler to the UE. The absolute grant sets the maximum power ratio which can be used for one or a group of UEs. E-RGCH, the relative grant channel, sends the relative grant which set grants increase or decrease the resource limitation compared to the previously used value. It has three values: 'UP', 'HOLD' or 'DOWN'. E-HICH is used by the Node-B HARQ to send the acknowledgments to the UE.

New UE categories are introduced in the Rel.'6 standard to support the enhanced uplink and listed in Table 2.6. Only QPSK is supported in this release. 16QAM is introduced in Rel.'7 which is not listed here.

2.5.1.2 Modified Protocol Architecture

Two new MAC entities are introduced in the UTRAN protocol to improve the scheduling on the uplink. MAC-e is added in the Node-B to support fast HARQ retransmissions, scheduling and

Table 2.6 HSUPA UE category (Rel.'7)

E-DCH category	Maximum number of E-DCH codes transmitted	Minimum spreading factor	Support for 10 and 2 ms TTI EDCH	Maximum number of bits of an E-DCH transport block transmitted within a 10 ms E-DCH TTI	Maximum number of bits of an E-DCH transport block transmitted within a 2 ms E-DCH TTI
Category 1	1	SF4	10 ms TTI only	7110	–
Category 2	2	SF4	10 ms and 2 ms TTI	14484	2798
Category 3	2	SF4	10 ms TTI only	14484	–
Category 4	2	SF2	10 ms and 2 ms TTI	20000	5772
Category 5	2	SF2	10 ms TTI only	20000	–
Category 6	4	SF2	10 ms and 2 ms TTI	20000	11484
Category 7	4	SF2	10 ms and 2 ms TTI	20000	22996

Note: When four codes are transmitted in parallel, two codes shall be transmitted with SF2 and two with SF4.
© 2008 3GPP

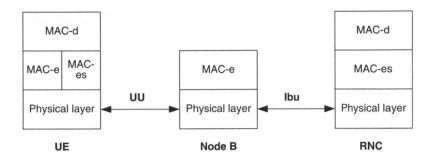

Figure 2.5   Enhanced uplink protocol architecture © 2008 3GPP

MAC-e demultiplexing. It resides in both the UE and Node-B. MAC-es, within the RNC and the UE, provides in-sequence delivery (reordering) and enables uplink soft combining for data when UE is involved in inter Node-B soft handover. Figure 2.5 shows the protocol stack of the enhanced uplink [3].

Since the transport channel (E-DCH) carrying HSUPA's user plane data is a dedicated channel similar to a R99 DCH, it is possible that soft combining can be supported for enhanced uplink. In the standard, the MAC layer functions on the network side are separated between the Node-B (MAC-e) and RNC (MAC-es) so that data frames from different Node-B for the same UE can be soft combined at the RNC level. Although soft combining provides diversity gain for HSUPA, it also requires that all Node-Bs involved reserve system resources for the same UE. It is a tradeoff between the capacity and performance that each network operator has to balance.

2.6 Summary

UMTS has been deployed quite widely around the world. UMTS is part of the GSM family of standards (GSM, GPRS, EDGE and future evolutions). Everything that makes up those technologies has been defined to the bit level with backwards compatibility and interoperability in mind; from switches to base stations to handsets. This level of standardization is unique to GSM and facilitates a level of interoperability between equipment from diverse manufacturers that is not possible in other wireless technologies. The standards and specifications for the GSM family of technologies are all developed in a group called Third Generation Partnership Project ('3GPP'), a collaboration that was started in 1998 in order to assure a common global set of GSM specifications.

In this chapter, we have provided an overview of a UMTS network's components and capabilities. The chapter introduces the reader to technology, architecture, nomenclature and performance. Since the book is focussed on the data capabilities of UMTS, namely HSDPA on the downlink and HSUPA on the uplink, we provided an overview of the channel structures, UE categories and features that need to be mastered prior to the successful planning, deployment and optimization of HSPA services. In the next chapters, we will dive deeper into RF planning and optimization for providing quality data services.

In a nutshell, these are the main benefits of the HSPA technology as compared to the legacy of Rel.'99 channels:

- higher peak rates (14.4 Mbps in downlink, 5.76 Mbps in uplink);
- faster network dynamics and reduced latency (90 ms with HSDPA, 50 ms with HSPA);
- new link adaptation and hybrid ARQ processes very efficient thanks to the shorter radio frames;
- the combined effect of all the improvements will result in higher spectral efficiency too.

References

[1] 3GPP Technical Specification 25.401, 'UTRAN Overall Description'.
[2] 3GPP Technical Specification 25.848, 'Physical Layer Aspects of UTRA High Speed Downlink Packet Access'.
[3] 3GPP Technical Specification 25.319, 'Enhanced uplink; Overall description; Stage 2'.

3

Applications and Quality of Service in HSPA Networks

In this chapter, we will discuss several topics related to the performance of internet applications in a mobile environment (HSPA networks in particular), and we will introduce some of the tools that operators have to control the end user service experience.

In Chapter 1 we presented current traffic trends in wireless networks, where data traffic is growing at a very fast pace and there is a multitude of integrated devices with more complex data capabilities. At points in this chapter, we will also highlight the different planning implications depending on the data business model targeted by the operator: typical handset user, smartphone users or PC users.

Since the data traffic is growing exponentially and each application has a different performance requirement, we will show that it is possible to improve the efficiency of the network if we are able to differentiate different types of traffic, and apply different priorities to each of them. This is the objective of the Quality of Service (QoS) concept that will be presented in detail in Section 3.2.

However, before plunging into the QoS capabilities of the network it is important to understand the traffic characteristics of various applications as this is the core of the concept of QoS. We will also compare how the wireless environment affects the performance of the internet applications as compared to the wired networks for which they were originally designed. Section 3.1 will review the parts of the network that can impact end user performance, and will show the increasing role of network latency in networks with high data rates. We will also discuss IP protocols and their impact on the overall performance because this protocol family has been found to be the root of many problems related to application experience in wireless networks. We conclude the chapter with a traffic analysis of the most typical applications in today's networks: web browsing, email and streaming. The analysis of how these applications work will help understand the underlying concepts for QoS: why traffic classes, priorities and delay requirements are created and what they mean. Points made

HSPA Performance and Evolution   Pablo Tapia, Jun Liu, Yasmin Karimli and Martin J. Feuerstein
© 2009 John Wiley & Sons Ltd.

throughout this chapter will help the engineers who design and optimize UMTS network save network resources while maximizing user experience.

3.1 Application Performance Requirements

When analyzing the performance of data applications it is important to consider the effect of all the network elements, from mobile to the data server. For instance, if we only look at the radio level we may find that the radio conditions are excellent, but maybe the link between the BTS and the RNC is congested and the total throughput is not as high as it could potentially be. Or we may find that even though all the elements are properly dimensioned, the network latency is so high that the TCP layer is prevented from achieving the maximum speeds. Figure 3.1 shows a HSPA network, indicating the major elements involved in a data transaction.

The end user experience on the data session will depend on the following elements:

- interference conditions (related to radio planning);
- capabilities of the BTS (RF and baseband capacity);
- available bandwidth on the Iub link (BTS-RNC);
- transport and baseband capacity of the RNC;
- available bandwidth on the Iu-PS link (RNC-SGSN);
- available bandwidth on the Gn link (SGSN-GGSN);
- transport and baseband capacity of SGSNs and GGSNs;
- core network layout (location of GGSNs and SGSNs);
- location and capacity of internal application servers (for intranet traffic);
- available bandwidth on the Gi link (GGSN-internet);
- external server location and capacity (for internet traffic);
- application and transport protocols.

There are fundamentally three reasons why the communication path in the network deteriorates:

1. Insufficient bandwidth.
2. Packet loss.
3. Delays.

These effects are typically linked to specific network elements: for instance, the transport backbone and core network elements may add delay or present bandwidth limitations, but will

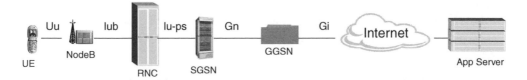

Figure 3.1   Network diagram for HSPA traffic (user plane)

seldom impact the packet loss. On the other hand, the radio interface may be affected by any one of these reasons, and certain tradeoffs can be established, for example packet loss can be improved if the number of retransmissions increase, which on the downside will hurt the delay. Within 2G and even early 3G wireless networks, the radio interface was typically the main bottleneck for the throughput and delay performance. However, with newly enhanced radio networks the air interface is highly efficient, putting more stress on other elements that were not taken into account before, such as the transport network or the core nodes. When bandwidth and packet loss are not the main problems anymore, network latency starts to play a role and needs to be carefully supervised.

3.1.1 The Role of Latency in End-user Performance

While in 2G networks the performance of the data session was typically assessed in terms of throughput, with the appearance of 3G networks, and especially HSPA networks, the key metric affecting the end user performance is the overall network latency. In networks with packet acknowledging schemes such as TCP/IP, the maximum effective data throughput is not necessarily equal to the system's peak rate: the latency can reduce the overall throughput due to the time required to acknowledge the data packets.

The fact that all packets being transmitted wait for an acknowledgment by the peer entity requires the presence of a packet queue on the transmitter (also called transmission window) where packets will sit waiting until they have been acknowledged by the receiver. Since there are physical or practical limitations to the size of the packets being transmitted (network Maximum Transfer Unit or MTU), and to the size of the transmission window, if the packet acknowledgment time – which is directly related to the system's latency – is too long, this queue will soon fill up and no more new packets will be able to be transmitted, thus effectively limiting the overall throughput. So, it may well be that our physical transmission media allows a very high speed, however, the practical speed will be reduced to a fraction of that bit rate if the network latency (measured by the packet Round Trip Time) is too high. For example, a 1 Gbps link with 100 ms RTT will only be able to deliver up to 100 Mbps. Thus, as peak bitrates increase, network latency needs to be reduced in order to benefit from the high capacity offered by the physical layer.

The effect of the latency can be observed in constant data transmissions such as FTP downloads, but is typically more noticeable in bursty, delay-sensitive applications such as Web Browsing. Figure 3.2 shows an example in which the same page is downloaded with different peak bitrates (from 300 kbps to 1 Mbps) and different round trip delays (from 150 ms to 300 ms). It can be observed that the impact of the peak bitrate is more noticeable when the latency is small, or in other words, large peak bitrates don't result in a better user experience when the latency is too large.

It is important to note that reductions in end-to-end delay could come from different parts of the network. For instance, a good planning and distribution of the GPRS core nodes can be very helpful in reducing network latency, especially for HSPA networks in which the RAN delay is significantly reduced. Therefore from a purely network performance and capacity point of

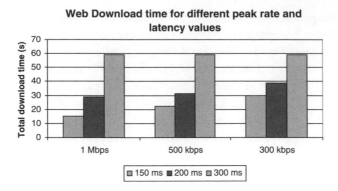

Figure 3.2 User experience of a web page download (CNN.com) as a function of peak bitrate and latency

view, it is desirable to have a distributed SGSN/GGSN network rather than concentrating them in gigantic packet switching centers.

3.1.2 Considerations of TCP/IP

The protocols involved in the communication can have a significant effect on the end user performance, since they can represent a source for inefficiencies (overhead) and delay. Figure 3.3 shows the typical protocol stack of data applications on a HSPA network (on the handset side).

In Figure 3.3, the first set of layers is specified within 3GPP and a basic description of their functionality was described in Chapter 2. The next two layers belong to the IP world and have been specified by the IETF (Internet Engineering Task Force).

The Internet Protocol (IP) was designed in the early 1970s for use in interconnected systems of packet-switched computer communication networks, to deliver a few basic services that everyone needed, services such as file transfer, electronic mail, remote logon, across a very

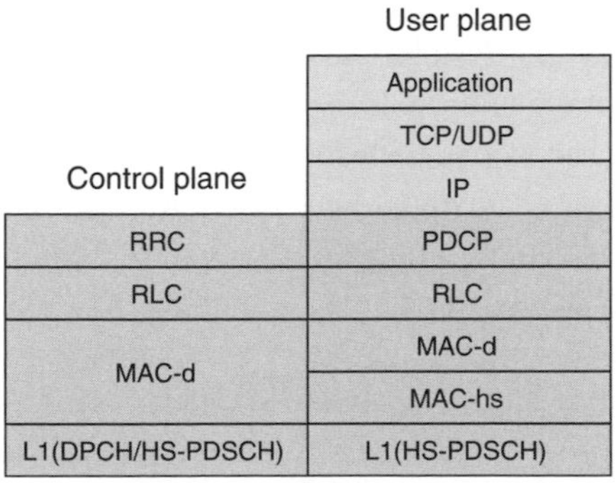

	User plane		
	Application		
	TCP/UDP		
Control plane	IP		
RRC	PDCP		
RLC	RLC		
MAC-d	MAC-d		
	MAC-hs		
L1(DPCH/HS-PDSCH)	L1(HS-PDSCH)		

Figure 3.3 UE Protocol in a HSPA network (DL only)

large number of client and server systems. The IP protocol suite has proven very flexible and almost 40 years later, it is still the most widely used protocol family in data communications, with enhanced functionality and new protocols being continuously standardized through the IETF forum.

3.1.2.1 IP Protocols

The IP reference protocol model consists of four layers: a data link layer, a network layer, a transport layer and an application layer. Out of these, the IP suite provides different protocols addressing the network and transport functions.

On the network layer, the IP protocol provides for transmitting blocks of data called datagrams from sources to destinations, where sources and destinations are hosts identified by fixed length addresses. The internet protocol also provides for fragmentation and reassembly of long datagrams, if necessary, for transmission through 'small packet' networks. There are several versions of IP protocols, the most well known are IPv4 and IPv6. IPv4 is the most utilized version of the protocol, while IPv6 is an evolution of the protocol that can satisfy the demand for the growing demand of IP addresses.

The IPv4 datagram consists of a header with variable size (between 20 and 60 bytes) and a data field. The header contains a variety of fields, including protocol information, checksum, and IP source and destination addresses, of 4 bytes each. The data field is also variable in size (between 0 and 65535).

The IP protocol family also provides a number of transport level protocols that are used for different types of traffic, depending on their requirements on data integrity, latency, etc. The most well known transport protocols are TCP and UDP: TCP (Transport Control Protocol) provides mechanisms for guaranteeing an ordered, error-free data delivery through a previously established data connection, while UDP (User Datagram Protocol) does not provide any guarantees but can provide a faster delivery of the data in a connectionless mode. In general terms, UDP is suitable for real-time or streaming applications, while TCP would be used for those transactions in which there is a need to fully preserve the original contents. Table 3.1 presents main differences between TCP and UDP.

The IP transport protocol provides a general framework for a larger set of networking applications that have also been standardized within the IETF: HTTP, FTP, TELNET, DNS,

Table 3.1 Main differences between TCP and UDP protocols

	TCP	UDP
Connection Establishment	3-ways handshake	Connectionless
Header size	20–60 bytes	8 bytes
Packet delivery	Ordered	Non guaranteed
Retransmissions	Cumulative and selective acknowledgments	None
Flow control	Sliding window	None
Congestion Control	Slow start, congestion avoidance, fast retransmit and fast recovery mechanisms	None

POP3, SMTP, etc. All the traffic flowing through the internet has a combination of the IETF transport and application protocols on top of the IP layer.

3.1.2.2 Considerations of the TCP/IP Protocol in Wireless Environments

Even though the IP protocol is used in a wide variety of data networks today, it is really not the optimum data transmission protocol for wireless networking. Most of these protocols were conceived to operate in wired data networks, or even through satellite links, but today's cellular networks can represent quite a challenging environment due to the rapidly changing radio conditions, which create sudden increases in error rates and can also result in a highly variable bitrate. Furthermore, the throughput in wireless networks is typically lower than those achieved in wired networking and the impacts from protocol overheads can be more remarkable.

Large TCP/UDP/IP headers can significantly reduce the network efficiency, especially for bursty applications with small packet sizes since the percentage of overhead compared with payload will be high. Let's look at a typical Voice over IP (VoIP) application as an example: for every packet worth of data the application needs to send 40 bytes of RTP/UPD/IP header, representing about 60% of the total data sent. The PDCP layer in UMTS can implement several header compression mechanisms which can reduce the IP overheads down from 40 to 1–4 bytes. Unfortunately, header compression is not available in the networks today although most vendors have it in their future roadmaps.

Connection oriented protocols such as TCP, which require packet acknowledgments from the receiver, can cause a great impact on the perceived data transfer if the underlying network presents high latency values. In the TCP protocol the effective transfer rate is calculated through the 'bandwidth-delay product', which indicates the effective amount of bytes that the network is capable of delivering at any given time. An elevated latency can cause a very poor experience in a very high bandwidth network, and thus both bandwidth and latency need to be taken care of. To ensure an efficient utilization of the network resources, the TCP Window Size should be adjusted to this bandwidth-delay product, which is 160 KB in the case of HSDPA, and 85 KB in the case of HSDPA+HSUPA. This means that to avoid significant rate reductions the TCP window size should be set to the maximum value of 64 KB. With this value it should be possible to achieve peak rates of 10 Mbps with HSPA.

Another consideration is the possible effect on latency due to IP segmentation in inter-mediate nodes. It is important to adapt the size of the packet itself to the capabilities of the network in order to reduce this effect. To achieve this goal, the Maximum Segment Size (MSS) should be adjusted to match the MTU (Maximum Transmission Unit), which defines the maximum packet size that the network is capable of processing. As an example, the MTU in conventional Ethernet networks is 1500 bytes, which in turn determines to a certain extent the MTU in many hybrid network systems since Ethernet is greatly extended. On the other hand, in modern gigabit Ethernet networks, the MTU can be up to 9000 bytes. There is always a tradeoff, though: while higher MSS/MTU can bring higher bandwidth efficiency, large packets can block a slow interface and increase the lag for further packets. As a result, in point to point links the MSS is usually decided at connect time based on the network to which both source and destinations are attached. Furthermore, TCP senders can use Path MTU discovery to infer the minimum MTU along the network path between the sender and receiver, and use this to dynamically adjust the MSS in order to avoid IP fragmentation within the network.

Due to the typical high latency values and limited bandwidth in wireless networks, the original sliding window acknowledgement scheme of TCP is not the optimum way to handle retransmissions in the cellular world, since it requires a complete retransmission of all the packets received after one error and would also provoke a data stalling while waiting to fill the receiver window. There are other acknowledgement mechanisms available in TCP, the most suitable for cellular being the 'Selective Acknowledgement' (SACK) in which the receiver explicitly lists what packets are acknowledged or unacknowledged. The SACK is widely implemented in today's protocol stacks.

Another consideration regarding TCP is the effect of the congestion control mechanisms implemented in the protocol. At the beginning of the TCP connection, or after every interruption of the communication (TCP timeout), the TCP protocol triggers a procedure called 'slow start'. This procedure resets and slowly adapts the receiving window of the TCP protocol, forcing more frequent acknowledgements on the packets following the initiation of the procedure. The TCP window size can only increase by 1 MSS per every packet acknowledged every round trip time (RTT). Since in cellular networks it is relatively frequent to have radio link losses, especially during cell changes, this procedure further degrades the end user experience. In recent years there has been some attempts made to alleviate this effect, and several versions of the TCP congestion avoidance procedure have been implemented. The 'TCP New Reno' version is the most suitable for wireless environment as it can operate with higher error rates and does not implement abrupt changes in the receiving window.

It is worth noting that the solutions to the above mentioned potential negative impacts of the IP protocols require a modification of the TCP/IP protocol stacks, or on the Operating System's parameters (MSS) both on the UE and on the application server.

3.1.3 Typical Application Profiles

The dimensioning of the different network elements is very much related to the type of applications the network will be handling. Different applications have different behavior and therefore have different requirements with regard to data loss, bandwidth and delay.

In this section we will present the main characteristics of three typical internet applications: web browsing (HTTP), email (SMTP, POP3/IMAP) and video streaming (RTSP/RTP). These applications present very different data transmission characteristics: web browsing is a non real-time, bursty application that requires the network to be responsive to user actions, email is non bursty and doesn't need to be so sensitive to user requirements, and video streaming is a continuous data stream with near real-time delay requirements. It is important to know these applications' network resource requirements because they need to be factored into network planning and optimization processes, as we will later present in Chapter 7.

3.1.3.1 Characteristics of a Web Browsing Session

Web browsing applications allow the user to transfer web pages from a server down onto their devices. The web page content is text mixed with multimedia objects like images, video, music, etc. The objects in the web pages can contain hyperlinks that connect them with other locations in the internet. The data transfer is controlled through an IETE protocol called Hypertext Transfer Protocol (HTTP) that is carried over a TCP/IP stack.

Let's analyze the behavior of the user, and what the network does at different steps of the communication process.

When browsing, first the user requests a specific internet address or URL (Uniform Resource Locator). From the end user's perspective, that triggers the download of a specific content that will be displayed in the screen during the following seconds, probably at different stages: first the text, then some images, maybe later some ad banners, etc. From the moment the user receives the text he/she will start to read the information and eventually after a certain period of time will either select a new URL on the address bar, or will click on a hyperlink in the downloaded page, which will trigger the download of a new web page. In engineering terms, the period during which the web page is downloaded is known as 'session', each of the elements that compose a web page are called 'objects', and the period between two sessions is called 'thinking time'.

On a lower level, these are the actions that will be taking place between the terminal and the network:

1. First, the user needs to be registered in the data network through the GPRS attach procedure. In many cases this step is not needed because the handsets are always attached (this depends on the handset configuration).
2. Then the device needs to obtain an IP address in case it didn't have one. This is achieved through the PDP context establishment procedure.
3. When there is IP connectivity, the web session can take place. The introduction of a web page address in the browser will trigger the establishment of a TCP connection needed to transport the HTTP protocol.
4. The URL needs to be translated into an IP address through a DNS lookup.
5. The HTTP protocol is ready to start and it will start requesting the different objects one by one until the page is fully downloaded. The reception of each and every one of these objects needs to be acknowledged by the device,[1] otherwise the web server will understand that they have not been properly received.

As demonstrated in Figure 3.4, multiple events occur even before the web page can be downloaded, each of them contributing to a longer web download experience: first, the GPRS connection needs to be established, then the DNS needs to resolve the server address;

---

[1] The HTTP protocol has been improved through several versions and the most extended nowadays is version 1.1. This version introduced the ability to request several objects from one page at the same time (pipelining), thus reducing the impact of network latency on the total download time.

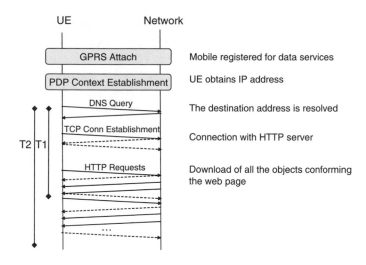

Figure 3.4   Generic diagram of a HTTP transaction on a UMTS network

afterwards a TCP connection will be established with the web server, and finally the server will send the client the different objects conforming the web page.

In reality, in data communication the GPRS Attach and PDP context establishment only take place at the very first session, so in communications with multiple sessions the importance of these delays are negligible. However, it can be significant for users that access their device to perform a single data transaction.

One important aspect to note about the web application requirements from the network is that the lower the latency, the faster the ack/nack process and therefore, the faster the web page gets downloaded. As shown in the example in Figure 3.2, a 2x reduction of latency results in almost 4x faster download speed.

In summary, the traffic pattern of web browsing applications will typically not be excessively demanding on network's peak rate requirements or average bandwidth due to their bursty and non real-time nature, therefore, the network dimensioning and planning could be more relaxed on data pipe capacity. However, the latency reduction is the main focus for web browsing applications. The end user will be expecting quick responses when visiting a new web address or clicking a web link. These expectations are intimately related to the overall network latency. In addition, the packet loss needs to be minimized to ensure content being delivered successfully to the end user. We will talk about how to achieve these goals in Section 3.2, where the relevant mechanisms for control of the Quality of Service (QoS) will be discussed.

3.1.3.2 Characteristics of an Email Transaction

The email application was one of the first applications introduced in the internet and since then it continues to be one of the most popular data applications. Due to the value it brings to corporations, it is also one of the most (if not the most) successful applications to be adopted by cellular data networks. The email application doesn't require a real-time response from the network. This makes it a very attractive service for operators to provide to their customer because of the low demand on network resources. There are two primary types of email protocols: POP3, which is more efficient but is less user friendly; and IMAP, which is more user friendly and less efficient.

The concept of providing email service is quite simple. The application allows sending messages from one computer, or device, to another one which may or may not be connected to the network at the particular moment that the message is sent. The message is simply stored in an intermediate server and then sent when the client is on-line. The original email applications only delivered plain text content, and since then the standard has been enhanced to support rich formats and file attachments as well. The text and attachment are sent together to the destination in a single and continuous transaction. The data stream pattern is similar to an FTP download (or upload).

The most popular email protocols are SMTP (Simple Mail Transfer Protocol) for uploading the message to the server, and POP3 (Post Office Protocol) or IMAP (Internet Message Access Protocol) for retrieving the messages from the server. The main difference between these two retrieving methods is that in the POP3 protocol the receiver polls the email server to check for new data, while in IMAP the server notifies the receiver of any changes in its account. In the POP3 case the connection with the server only lasts as long as the email download, while in the case of IMAP the client can be continuously connected to the server.

Although POP3 is the most used email retrieving protocol, IMAP is the most adopted for corporate email applications due to the faster response, possibility of message tagging and other advantages. Both Microsoft Exchange and IBM Lotus Notes implement proprietary versions of the IMAP protocol. The main disadvantage of IMAP is the high consumption of resources as compared to POP3.

The end-user expectations with regards to email is to experience a relatively fast upload to the server once the 'send' button is clicked, and get a relatively fast response when the 'receive' button is pressed. In the IMAP case, since there is no need to click the receive button, the 'receive' experience is masked to the user. In any case, there is no need to grant any constant bitrate or very fast response for email application. For these reasons email is typically considered a 'background' service.

3.1.3.3 Network Impacts of Near Real-time Video Streaming

With the increased speed of the internet links at home, video and audio streaming have become extremely popular applications. Most TV and radio channels have an option to broadcast their contents through the internet. Consumers are more and more used to streaming services, such as videoconferencing, online video rentals, 'youtube' type of services, etc. With the data enhancements provided by advanced 3G network, streaming services are also becoming increasingly popular in the wireless world.

Streaming services can deliver audio or video contents (or both). The streaming traffic contains data that has been previously encoded at a certain rate (frames per second), and therefore that data needs to be received at the client terminal at the same specific rate. Typically, the coding of the streaming frames has a 'historic' component, meaning that one frame is coded as a function of the previous frame that has been sent. This implies a potential for losing several frames in a row when one of the frames is lost. On the other hand, the client can reproduce the content even when not all the media frames are properly received: if the data loss is small, the user will hardly notice any problem, however with a high packet loss the streaming media will present interruptions.

The streaming protocols were originally created by the International Telecommunication Union (ITU-T) with the design of the H.320 and H.323 protocols. Today there are many protocols specifically designed for streaming media, like RTSP, RTCP, RTP, Adobe's RTMP, RealNetworks RDT, etc. and other generic protocols can be used as well, like HTTP/TCP or UDP.

The UDP protocol, and those based on UDP like RTP, is more suitable for real-time data since it's a simpler mechanism with less overhead and lower latency. On the downside, it's unreliable and doesn't guarantee an ordered delivery so it can lead to errors and interruptions during the playback. Also, many firewalls block UDP protocol because it's less secure, which forces the use of TCP in streaming protocols.

Streaming protocols built on top of TCP guarantee a correct delivery of the content; however this is achieved through retransmissions and can cause timeouts and data stalling. In order to minimize this effect, streaming clients usually implement a buffer that stores a few seconds of the media content in the client before it starts to display it.

In addition to the transport protocol, there is a streaming control protocol that implements commands such as 'setup', 'play', 'pause', etc. There are several of these protocols available, the most well known being RTSP, RTCP and SIP (Session Initiation Protocol). SIP has been gaining momentum within the cellular industry since it is the standard used by the IP Multimedia Subsystem or IMS. These control protocols are generally combined with the RTP protocol.

The streaming traffic flow typically presents a constant bitrate transmission from source to client, with frequent traffic peaks or valleys which are caused by the encoding mechanism. Due to the fluctuations on data networks, clients usually implement a de-jitter buffer which introduces a certain delay at the beginning of the media reproduction. This ensures a smooth playback compensating for differences in delay arrival between packets or temporary packet loss. With TCP clients, a temporary data loss will result in a higher data transfer in the periods after that event since the de-jitter buffer needs to be refilled. This effect is known as re-buffering. With UDP the client would just perceive degradation on the quality of the images or sound due to the lost frames. Figure 3.5 shows an example of video streaming over TCP from the internet (CNN.com video). It can be observed that in that particular session the instant bitrate was usually around 300 kbps, with constant fluctuation and occasional peaks due to re-buffering events.

When offering streaming applications, network operators need to be ready to reserve the appropriate bandwidth to offer a quality service. Streaming traffic is very resource intensive and for a good customer experience the applications require a certain guaranteed transmission bandwidth. The specific network requirements need to be determined by the operator. For

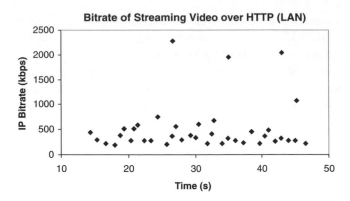

Figure 3.5   Streaming bitrate capture from CNN.com video over LAN

instance, an operator planning to offer video streaming on handsets could provide guaranteed bitrate as low as 128 kbps per connection. However, if the same service is to be offered on laptops, the operator would have to guarantee a minimum of 300 kbps per connection.

Now that we have explained the characteristics of typical applications, the next section (Section 3.2) will present the mechanisms that 3GPP offers to control these applications' Quality of Service (QoS) requirements.

3.2 Support of QoS in HSPA Networks

In the previous sections we have analyzed how the data traffic is increasing in the cellular environment, and demonstrated the different resource requirements of various applications. This traffic mix will become more complex as more applications are made available, especially those bundled with real-time communication features. Since the traffic is growing and the different applications being used have different performance requirements, it's possible to improve the efficiency of the network by applying special treatment to each traffic type. This is the objective of the Quality of Service (QoS) concept.

As an example, consider a link that serves two users: one is using a video streaming service and another one is downloading a heavy email. Let's also assume that the link has a limited bandwidth and cannot serve both users simultaneously as good as it would do it if both users were doing their transactions at different times. Figure 3.6 illustrates the situations in which all the network resources are available for each user (left), and when both users are sharing the link, first without any kind of QoS mechanism (middle) and then with certain QoS treatment (right).

Under the scenario presented in this example, when both users share the link and there is no Quality of Service control, both applications will be treated similarly. Both users' data experience would be affected since the email cannot be downloaded as fast as in the first case, and the video streaming user will certainly experience interruptions during the playback. On the other hand, as we reviewed in the previous section, the customer using an email

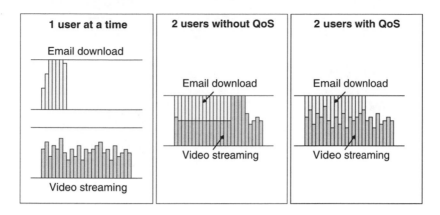

Figure 3.6 Link traffic example at different conditions: separate users (left), simultaneous users without QoS (middle) and simultaneous users with QoS (right)

application doesn't really notice that the download takes a couple more seconds to finalize. However any interruption during video playback will be noticed by the customer using video streaming. In this case, it could be possible to serve both users satisfactorily by assigning a higher priority to the streaming traffic, and leaving the remainder of the bandwidth for the email user. Thus, the email traffic will adapt to the varying bitrate of the streaming, transmitting less traffic during the peaks and filling the gaps in the other cases. The result is a similar email download time than in a fully equal distribution of resources; however the video user this time will be fully satisfied. The beauty of the QoS control is that it can help improve the network efficiency without impacting the user perception.

The QoS concept is applicable to any data network, however it really only makes sense when the network resources are limited. Typically in wired data networks there are no critical bandwidth limitations as it is the case with cellular networks where spectrum and backhaul are scarce and expensive resources. This made it a perfect ground for the development of QoS mechanisms. Today, both IP and 3GPP networks have QoS control mechanisms defined. Although they share some similarities, they are quite different in many aspects.

3.2.1 3GPP QoS Attributes

3GPP QoS was standardized first in Rel'97/98 (GPRS), and since then has gone through significant changes. Rel'99 introduced the QoS classes and attributes as we know them today, although it was mostly focussed on the Radio Access. In Rel'6 major changes were introduced to provide an end-to-end view to the process, including interaction with IP QoS, although the concepts were very concentrated around IMS services. With Rel'7 new enhancements were added to contemplate a more generic service environment and with a stronger focus on IP connectivity. The main specifications covering QoS aspects can be found in [1–3].

Since the GPRS core architecture is the same for GPRS and UMTS, the QoS mechanisms are common to both technologies. In practice, while the networks have been ready for QoS to a

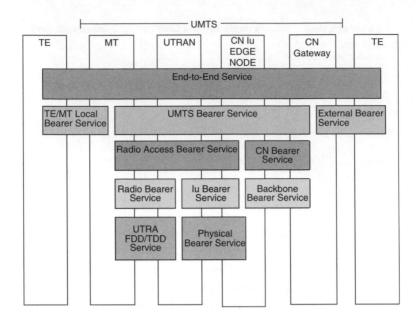

Figure 3.7   UMTS QoS entities since Rel'99 [1] © 2008 3GPP

certain degree, it is rare to find today's networks that apply these procedures since the handsets have usually lagged behind and typically don't request any special treatment for their applications. It is, however, very likely that QoS will gain momentum in the near future due to the awakening of wireless data usage, and the availability of a mature QoS architecture like the one finalized with Rel'7.

Figure 3.7 illustrates the different components of the 3GPP QoS architecture [1]. Although the architecture has evolved through the different 3GPP releases, this diagram applies to all releases from Rel'99 to Rel'8. Each of the different entities in the diagram is responsible for QoS treatment in their particular communication link, with the system providing the appropriate coordination and mapping between different entities to ensure a consistent treatment from end to end. Precisely the improvements in the different QoS releases are mostly focused on smoothing this integration, introducing the coordination with external networks such as IP.

The QoS mechanisms defined in 3GPP cover two different aspects of traffic differentiation: user and service differentiation. In the case of user differentiation, the subscribers may pay for different data priority levels and this will be reflected on how fast they receive or transmit data as compared to other users. Also, users may subscribe to receiving certain traffic type such as streaming, which could otherwise be blocked by the network due to the high resource consumption involved. In the case of service differentiation, the traffic generated by the applications will be classified into different profiles which will be assigned a specific QoS treatment suitable for that particular data type. While the first approach provides the operator with a new tool for additional revenue, the second approach provides a more efficient way to utilize the network resources. These methods are not mutually exclusive and they can be combined in the same network.

To enable traffic differentiation, the standards have introduced a series of parameters that can be specified differently for different users or traffic profiles. Each user has a 'subscribed QoS' profile stored in the Home Location Register (HLR) which contains the maximum QoS requirements they can request to the network. During the communication, handsets will also request a specific QoS profile, which will depend on the application type; the requested profile will be contrasted with the subscribed profile, and the most restrictive profile will be used. The most relevant attributes configuring a QoS profile are [1]:

- **Traffic Class**. Identifies how to treat a particular communication. There are four traffic classes: Background, Interactive, Streaming and Conversational.
- **Traffic Handling Priority (THP)**. Permits priorities to be assigned to different users/ application types within a particular traffic class.
- **Allocation/Retention Priority (ARP)**. Indicates whether a service can steal resources that have already been assigned to other user, and vice versa.
- **Bitrate**. The required speed (kbps) for the connection. Bitrate limits can be specified either as *Maximum Bitrate* or *Guaranteed Bitrate* (for streaming and conversational communications).
- **Bit/Packet Loss**. Specifies the maximum transmission loss that the application can support without too much degradation. It can be defined at bit and packet level.
- **Transfer Delay**. Represents a boundary to the packet transfer time and can also be used to control jitter.

The following table (Table 3.2) illustrates the typical QoS requirements for some example application profiles, including traffic class, packet loss, delay and bitrate.

It is up to the wireless network operators to decide what QoS parameters are suitable for each type of application. For instance, a real time application such as Video Conference over IP should use the Conversational traffic class with a guaranteed bitrate according to the source (e.g. 64 kbps) and a limited maximum delay (e.g. 300 ms), and could afford a certain packet loss (e.g. 5%). On the other hand, a push email service can be set to the Background traffic class, which has no bitrate guarantees or delay requirements, but the contents should be preserved, thus a 0% packet loss is required at the application level.

3.2.2 Negotiation of QoS Attributes

Before a data transaction begins, handset and network will engage in a negotiation process that will determine the QoS attributes that better suit the user needs, considering the available network resources. It is important to note that since QoS requirements are based on application type, it is typically the handset that should request a specific treatment for its traffic. Nowadays, however, while the networks are usually ready for QoS, most of the handsets fail to link the application requirements to the QoS in the lower layers and the communication takes place without any specific QoS differentiation.

Table 3.2 Example of application types and their corresponding QoS attributes

	FTP/Email download	HTTP/WAP, Online games, Blackberry	Video/audio broadcasts	Voice over IP, Videoconference
Traffic Class	Background	Interactive	Streaming	Conversational
Typical traffic shape	Large amounts of data downloaded in a single (long) session	Bursty data with think times. Typically small amounts of data	Continuous data stream at a constant bitrate with certain bitrate fluctuation	Continuous data stream at a constant bitrate with certain bitrate fluctuation
Symmetry	Asymmetric	Asymmetric	Asymmetric	Symmetric
Content preservation	100% error free	100% error free	Can support data loss	Can support data loss
Latency requirements	Tolerates delays and delay variations (jitter)	Delay aware, although not very sensitive. Tolerates jitter	Delay sensitive, can tolerate a certain jitter (due to buffering)	Delay stringent, very sensitive to jitter
Bandwidth requirements	No special requirements	No special requirements	Minimum bandwidth guarantee	Minimum bandwidth guarantee

The negotiation of QoS attributes is performed during the establishment of the Packet Data Protocol (PDP) Context. The PDP context is created when the data user needs to establish a connection (for instance when an application opens a 'socket'). At the moment of the PDP establishment the UE will receive an IP address and a certain number of attributes to be used for that connection. Amongst them are the QoS attributes. The PDP context is established between the UE and the GGSN, however it is not accessible from the Radio layers, and therefore the final parameters need to be communicated to the RAN. This is done at the moment of the establishment of the Radio Access Bearer (RAB) and hence on the underlying Radio Bearers (RB) and Iu Bearer (see Figure 3.7 for details on the bearer's architecture).

As previously indicated, in an ideal case the QoS parameters should be requested by the handset, depending on the application that is requesting a data tunnel. From the handset side, there are two ways to request for QoS: the easier one is to have different applications access separate GPRS Access Points (APNs). In this case, the handset actually does not request a set of QoS attributes. The network will apply a different set of parameters depending on the APN that has been accessed. The better method is for the handset to request the specific QoS parameters in the PDP context creation command. In such a case there is no need for the operator to offer multiple APNs. From the network side, there is a third way to achieve QoS differentiation. It doesn't require the handset to request for a specific QoS: the network (GGSN) may have intelligent functionality to determine the requirements based on the traffic flowing through it, and may subsequently ask the UE to modify the existing PDP context. The drawback of this kind of solution is that it will not support multiple data profiles from a single terminal, and would also create too much signaling load.

Once established, the PDP context can only be applied to the negotiated QoS profile. If the data profile changes during the communication, the PDP context can be renegotiated, or alternatively a second PDP context may be established. When establishing multiple PDP contexts it is preferable to create one primary, and then several secondary contexts, since each primary PDP context will consume one public IP address from the operator's pool.

The PDP attributes need to be further communicated to the different elements in the network in order to shape the traffic according to the QoS settings. Figure 3.8 shows a network diagram indicating what QoS information is known at every network node, and the specific functionality they implement for QoS.

In a 3GPP network, the following elements are required in order to properly manage the QoS for the different users. Their functionality will be reviewed in detail in the next chapter (Chapter 4):

- **Admission Control** on the air interface and on the last mile transport network. This functional element decides whether the connection is permitted or not based on existing resource utilization and overall quality.
- **Resource Manager** governs distribution of channel resources including data and control traffic.
- **Congestion Control** takes action in case of overload based on user profile.

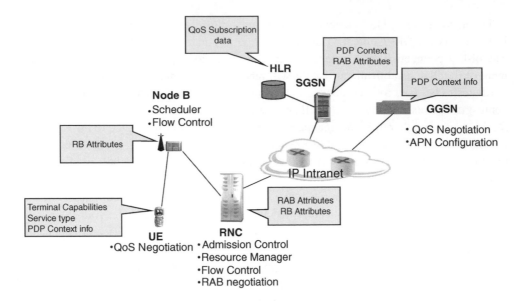

Figure 3.8   Network diagram of QoS functions and information (Rel'4)

- **Flow Control** controls packet scheduling over the transport network.
- **Air-interface Packet Scheduling** ensures users and bearers get the QoS they have requested and paid for.

The required QoS parameters are known within the RAN nodes through the Radio Access Bearer attributes, however the Radio Access Bearer is terminated at the RNC and with the specified architecture it was impossible to propagate these values to the NodeBs. This architecture works fine for Rel'99 channels, since all the required radio resource management elements are physically located in the RNC, however it presents a limitation with respect to HSDPA and HSUPA channels, since in these cases the packet scheduling functionality is located in the NodeB.

3.2.3 QoS Modification for HSPA

With the introduction of HSDPA, the NodeB gains more control over the communication beyond the pure physical layer, and it is therefore necessary to ensure that the proper QoS treatment is guaranteed. Since the existing QoS architecture was terminated in the RNC, the standard had to introduce some changes in Rel'5 to make sure that the QoS parameters were properly propagated down to the NodeB once determined at the RAB setup.

The main changes introduced by HSDPA are captured in [4] and affect the interface between the NodeB and RNC, called NBAP. These changes include the creation of new QoS parameters specifically designed for HSDPA and HSUPA. The following picture illustrates the flow of the QoS configuration parameters (Figure 3.9).

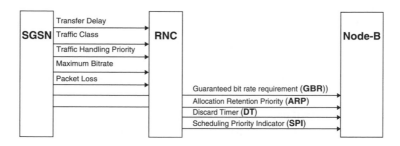

Figure 3.9   QoS parameters known at the RNC and NodeB levels

The parameters controlling the Quality of Service at the NodeB are four, two of which can be mapped directly from the RAB QoS parameters: the Guaranteed Bitrate (GBR) and the Allocation and Retention Priority (ARP). The other two parameters are new, although their concept is similar to other existing parameters.

- The Scheduling Priority Indicator (SPI) provides up to 16 different values that can be used by the NodeB to prioritize different traffic flows. In concept, it is similar to the Traffic Class and Traffic Handling priority. The combination of TC, THP and ARP provide up to 18 different values which could almost be mapped one-to-one to different values of SPIs.
- The Discard Timer (DT) indicates the maximum time that the packets from one connection should be stored in the NodeB queue. This information is inherently related to the transfer delay and can be used as input for a time-sensitive scheduler, or as a method to discard packets when they become too old.

Although these parameters are defined in the specifications, it is up to the vendors how to use them and for that reason the real effect of those parameters depends on the implementation.

A major drawback for introducing this modification is the need to replicate the QoS configuration, since the system will now require a different set of parameters for Rel'99 channels and another set for HSPA channels. On the other hand, the modification has introduced an advantage to the configuration of packet services since now all the HSPA QoS parameters can be applicable to any type of traffic type. Now it is possible for an operator to define a Guaranteed Bitrate for an interactive user, for instance, without having to treat the user as a streaming type.

These changes introduced in Rel'5 permit that the main RAN and core nodes of the HSPA network are able to consistently propagate and apply the QoS profile required by the application. However, in order to fully achieve a consistent QoS treatment across the network there are other factors that need to be considered, such as the translation of these attributes within elements that have not been specified by 3GPP, for instance, IP routers. These enhancements are part of 3GPP Release 7, which will be further discussed in the architecture evolution part of Chapter 8 (HSPA Evolution).

3.3 Summary

The following are the main takeaways from this chapter:

- Data traffic is increasing significantly and operators must be ready to plan their networks accordingly.
- The network planning will depend on the operator strategy with regards to device offering (handheld, smartphone, PC data card) and service offering (best-effort, real-time).
- The quality of the different services should be analyzed end-to-end, considering all possible bottlenecks in the communications path.
- Latency is a key performance metric in HSPA networks because it greatly affects end user experience.
- TCP/IP can have a significant effect on throughput and latency. TCP parameters should be carefully reviewed in handsets and application servers.
- Protection of different aspects of applications' performance can be achieved through appropriate Quality of Service functionality.
- To ensure an efficient utilization of the resources the operator needs to properly deploy an end-to-end QoS network, including – most importantly – the handset support.

References

[1] 3GPP Technical Specification 23.107 'Quality of Service (QoS) Concept and Architecture'.
[2] 3GPP Technical Specification 23.207 'End to End Quality of Service (QoS) Concept and Architecture'.
[3] 3GPP Technical Specification 23.802, 'Architectural Enhancements for end-to-end QoS'.
[4] 3GPP Technical Specification 25.433, 'UTRAN Iub interface Node B Application Part (NBAP) signaling'.

4

Radio Resource Management in UMTS/HSPA Networks

To be efficient and competitive, wireless carriers must design and dimension their networks to create a balance between service quality and the resources needed to serve the traffic. Resources, such as network processing elements and backhaul, are limited either because they are expensive (e.g., site developments, RF or baseband modules, leased lines, etc.) or because the resources themselves are scarce (e.g., spectrum and transmit power). In order to achieve maximum efficiency, network resources should be allocated with changing network conditions such as traffic load, radio conditions and application type.

In a packet data system with highly variable traffic, where large demand changes exist over time and location, resource allocation flexibility is the key. Rigid 'nailed up' assignments of timeslots, power and codes are extremely wasteful. Instead, dynamic and flexible sharing of resources unlocks both high throughputs and capacities. However, tradeoffs between voice and data capacity and quality need to be made – nothing comes for free. This requires careful selection of resource allocation algorithms and parameter settings.

One of the technical realizations in developing efficient packet switched technologies such as HSPA was that one big pipe was better than many smaller pipes. This is the underlying principle in the evolution of CDMA technologies such as CDMA2k and UMTS into 1xEVDO and HSDPA. One big pipe shared amongst many users allows the full capacity of the channel to be dedicated to a single user if needed. The idea is to be able to burst data out to users when they are in good conditions (e.g.: high SINR ratios) while intelligently allocate the bandwidth to multiple users with varying radio conditions to achieve optimum capacity and throughput. The function of sharing resources efficiently between different users is called Radio Resource Management (RRM). Although RRM is an important part of the WCDMA system, the actual implementation by different infrastructure vendor varies.

RRM functions typically include support for Quality of Service (QoS), enabling the ability to transport more data to users or applications with higher priorities. QoS treatment also permits

HSPA Performance and Evolution   Pablo Tapia, Jun Liu, Yasmin Karimli and Martin J. Feuerstein
© 2009 John Wiley & Sons Ltd.

rich application profiles powered by the high speed data capability of the 3G network to compete for the resources with conventional services such as circuit-switched voice. To effectively share the HSPA big fat pipe among many users, there are numerous tools in the operator's bag of tricks, but we can divide them into two main categories: (a) cell level capacity enhancing algorithms; and (b) link level (i.e., per user level) performance enhancing algorithms.

In 2G E-GPRS and 3G Rel.'99, most of the radio resource management functionality was performed in the control nodes (BSCs or RNCs), whereas with HSPA implementation this functionality has been primarily pushed down to the NodeB. Having the RRM functionality resident in the NodeB enables faster responses to changing radio conditions and results in improved overall capacity and quality.

In this chapter we discuss the RRM functions for High Speed Data Services (HSDPA and HSUPA). The key cell level RRM algorithms for HSPA are the following:

- **Admission and congestion control:** maintains quality for the existing users and efficiently assigns resources to new users;
- **Packet scheduler:** controls the distribution of the resources in the cell amongst all active users. Assigns each timeslot to a single user or a group of users (called 'code multiplexing') based on radio conditions and quality of service requirements for each application; and
- **Power allocation:** controls the relative allocation of transmit power between voice and HSDPA communications.

However, cell level mechanisms to ensure quality do not provide a complete solution to end-user quality of service. Operators are able to fine-tune user performance on a connection-level as well. Connection-level quality-enhancing algorithms are the following:

- **Mobility management:** NodeB controlling the HSPA mobility such as cell transitions while maintaining high user throughput; and
- **Power control and link adaptation:** control the call quality for each individual user while not wasting resources that can be used by other users.

Figure 4.1 depicts the nodes in the network where each of the RRM algorithms resides.

In this chapter, the focus is on introducing and describing the fundamental concepts of RRM. Later, Chapter 5 reviews the essential parameters controlling the RRM functions and provides recommendations on their optimum settings. Chapter 6 presents our extensive lab and field test results on the performance gains of many of the RRM algorithms under real network conditions.

4.1 Admission and Congestion Control

Admission control (AC) and congestion control (CC) are the two major mechanisms governing the assignment of radio resources (power and channelization codes) in a UMTS cell. Admission control is important because it maintains system stability by blocking incoming

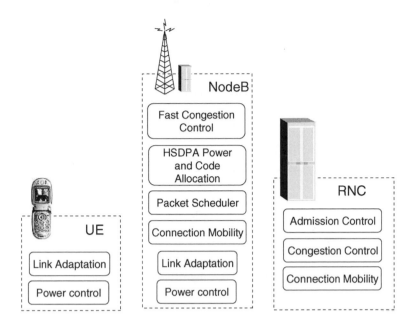

Figure 4.1 Block diagram of the HSPA network elements identifying the locations of the various RRM algorithms

calls when resources are insufficient. Congestion control, on the other hand, protects the overall quality of the cell by downgrading or even dropping low priority calls.

To understand these algorithms, it is helpful to review the basics of CDMA technology. In CDMA networks all the users in the cell and in the adjacent cells are sharing the same carrier frequency, which is called unity frequency reuse. Since all connections share the same frequency, an increase in the interference level will affect all users in the geographical area. With spread spectrum systems, it is possible to guarantee the quality of the connections as long as the interference is maintained below a certain threshold, which depends on the network and service characteristics. On the other hand, as Figure 4.2 illustrates, if the overall interference level increases beyond the stable region, crossing the threshold into overload, then the system becomes unstable. This causes many users in the cell to suffer serious quality degradations from interference. This could happen when a cell operating near the overload state admits new users, or when the cell assigns too much power to existing users. The functions of the AC and CC units are to ensure that the cell doesn't operate beyond this stability limit. More detailed information on these basic UMTS concepts can be found in [1].

In general, the AC and CC functions manage four types of resources of the RAN:

- transmit power;
- channelization codes;
- baseband resources (channel elements);
- Iub transport bandwidth.

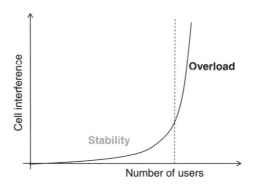

Figure 4.2 Operating load curve of a CDMA system showing stable and overload (unstable) regions versus the traffic load (number of users)

The following subsections explain how the algorithms manage each of these four types of resources.

4.1.1 Management of Transmit Power Resources

One goal of the AC and CC functions is to regulate the total powers for the uplinks and downlinks. By controlling the powers transmitted from or received by the NodeB, the network interference levels can be minimized.

There are two primary methods for power resource management:

1. Based on thresholds established for the total transmit power on the downlink and received total wideband power (RTWP) on the uplink.
2. Based on an 'equivalent user' capacity number, in which the operator defines the specific number of users that should be admitted in the system.

In the second case, the typical power consumptions of specific traffic types (e.g. voice AMR 12.2) are defined as the 'per user power consumption'. Defining the number of users that a cell can support indirectly sets the power limit for the AC and CC functions. Both methods are comparable since the 'equivalent user' number will ultimately be translated into a power threshold. Therefore in the rest of this section reference will be made only to the first method based on power thresholds.

Figure 4.3 illustrates how the AC and CC mechanisms jointly manage the cell's transmit and receive power resources. As shown in the figure, AC starts working when the total power reaches a user-specified AC threshold, which is lower than the CC threshold. If the power level in the cell continues increasing above the AC threshold, then CC takes over. CC employs more drastic measures to prevent the cell from entering the overload region, which could dramatically degrade communications in the cell.

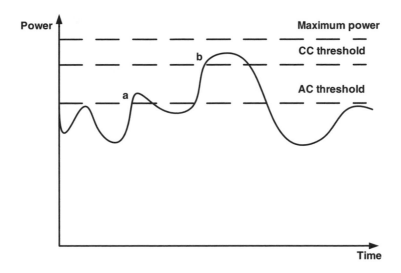

Figure 4.3   Illustration of power resource management using the AC and CC mechanisms in RRM

After the total power exceeds the AC threshold, then new calls originated or terminated in this cell will be rejected; however, calls requesting soft handover from other cells can still be admitted. Conceptually, all UMTS AC and CC algorithms work in a similar manner; however, the specific triggering parameters depend on the vendor implementations. For instance, some vendors may use an absolute CC threshold as illustrated in Figure 4.3, while others may use hysteresis on top of the AC threshold. Some vendors may use a different power threshold for accepting soft handovers.

CC can be triggered based on the following conditions:

- events or
- when the power increases beyond a specified threshold, as described above.

Examples of event triggering are rejection of incoming calls or rejection of a handover request. It is usually possible to configure the number of events that will trigger the CC function. Threshold triggering, on the other hand, is more flexible since the threshold parameter can be configured differently according to the actual network condition and traffic profile.

When the CC function is triggered the system needs to determine which service is the best to be downgraded or released to reduce the overall interference level. The decisions on the services to be downgraded are based on the QoS priorities of existing services, and the overall resources consumed by each particular user. For instance, non real-time services will have lower priority than real-time services; background services will have the lowest priority among all services. Additionally, the service which is using the most system resources should typically be downgraded first.

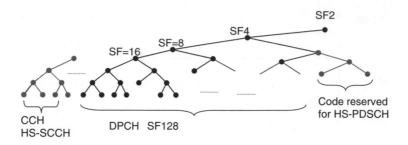

Figure 4.4   Code tree illustrating spreading factors (SF) and code usage in a WCDMA system

4.1.2 Management of Channelization Codes

In addition to power, the AC and CC functions also need to manage the availability of physical transport channels. For example, AC can reject incoming calls if there are no channelization codes available for it. On the other hand, CC can identify code shortage situations and trigger specific mechanisms that will help free the necessary resources.

The channelization codes used in the WCDMA system are orthogonal variable spreading factor (OVSF) codes. Figure 4.4 shows one example of realization of the code tree and assignments. In principle, the number of available codes is directly related to the spreading factor (SF) the channel is using. For instance, the spreading factor for the Rel.'99 Dedicated Channel (DCH) is 128; therefore the maximum number of available OVSF codes is 128. In the case of HSDPA – which uses SF of 16 – although there are 16 channelization codes available on the code tree, only 15 of them can be used by the traffic channels because the remaining one is used for signaling.

The code resource management algorithms depend on the vendor implementations. Because the SFs for different channels can be different, a well designed code assignment algorithm is needed to allocate the codes efficiently. When voice and HSDPA are both supported on the same carrier, code allocation is much more complex. It may require that codes assigned to different channels be rearranged (packed) to make room for the new channels.

4.2 Packet Scheduler

4.2.1 HSDPA Scheduling

The goal of the HSDPA scheduler is to maximize the spectrum efficiency of the cell while maintaining the QoS requirements for different data services. As discussed in Chapter 2, the HSDPA scheduler resides in the NodeB. It is a fast scheduler which can make decisions on per TTI (2 ms) basis. Although the implementation of the scheduler depends on the vendor, there are three fundamental types of algorithms typically used:

- Round Robin;
- Maximum C/I; and
- Proportional Fair.

The Proportional Fair algorithm is the most accepted because it achieves a tradeoff between cell capacity and user quality. In this chapter, we explain the algorithms and benefits of each scheduler type, along with recommendations for usage. Chapter 6 presents supporting lab and field data comparing the performance of these scheduler algorithms in real-world scenarios.

In general, the HSDPA scheduler typically assigns transmission turns to different users based on any (or all) of the following criterion:

- channel condition, communicated by the UE through the CQI report;
- QoS parameters, such as the traffic class, traffic handling priority (THP), allocation and retention priority (ARP), scheduling priority indicator (SPI) etc.;
- UE traffic control specifics, such as the time data has been in the buffer or the amount of data in the buffer; and
- available system resources, such as transmit power, codes and Iub bandwidth.

The support for Quality of Service requires that the HSDPA scheduler consider both the traffic class (Conversational, Streaming, Interactive and Background) and the QoS attributes (priorities, delay requirements, etc.). For instance, conversational bearers with stringent delay budgets should be treated with the highest priority, while background bearers – the least demanding for timely delivery of data packet – can be assigned a lower priority. This sometimes requires that the packet scheduler ignore the measurement feedback CQI provided by mobile terminals and deliver the data to users with less favorable channel conditions.

The scheduler decides how to serve each and every one of the users of the sector by assigning the following resources:

- **Scheduling turns:** in every TTI (2 ms) the scheduler decides which user can transmit and which cannot.
- **Number of codes:** if several users are transmitting on the same TTI the scheduler will decide how many parallel codes are assigned to each of them.
- **Power per code:** if HSDPA codes are assigned to different users, the scheduler may decide to assign a different power to each user, based on the available HSDPA power and the specific user requirements.

Figure 4.5 illustrates different scheduling techniques that could be implemented, using a simplified example. The diagram shows three HSDPA users, with User 1 having lower throughput requirements than Users 2 and 3. The diagram on the left illustrates the case where all the HSDPA codes are assigned to one user at a time, while the right diagram introduces one more degree of complexity by multiplexing users in every TTI.

The following sections provide details on the different types of packet scheduler algorithms.

4.2.1.1 Round Robin Scheduler

The Round Robin (RR) packet scheduler algorithm is the simplest one. It distributes the scheduling transmission turns equally among all active HSDPA users, regardless of the radio

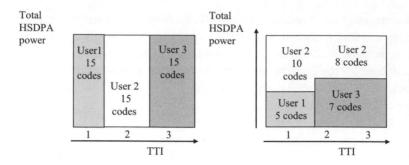

Figure 4.5   Example of different HSDPA scheduling strategies

channel condition and the QoS requirements of the application running on the mobile devices.
Figure 4.6 shows an example of how RR works in a system without code multiplexing. It can be
observed that users in poor radio conditions (such as User 4, far from the cell in the diagram)
receive the exact same number of turns as users in perfect radio conditions (such as User 3 close
to the cell). As a result of this, the overall spectrum efficiency using the RR algorithm is not
maximized. The fairness in time sharing of the system resource creates unfairness to those UEs
which are under good radio conditions and starving for throughput.

4.2.1.2 Max C/I Scheduler

The Maximum C/I algorithm, on the other hand, aims to optimize the overall spectrum
efficiency of the system by allocating the radio resource to the UEs with the best channel

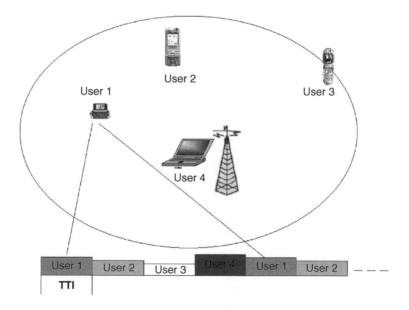

Figure 4.6   HSDPA Round Robin scheduler example

conditions. Naturally, the instantaneous channel condition of each user can vary independently, and the fast scheduling capability of HSDPA provides the opportunity for the system to transmit on channels with the best radio condition at any time. For most types of radio environments, such as macro cells, the correlation among different radio channels is low, and the HSDPA scheduler can take advantage of the channel diversity, therefore increasing the overall system capacity.

A system running this algorithm has the best overall cell throughput compared to the other methods; however, it creates unfairness to UEs under poor radio conditions, especially those at the cell edge, which may potentially get zero or very low throughput. One can see that maximal overall throughput for the cell does not translate into equal user experiences.

4.2.1.3 Proportional Fair Scheduler

The Proportional Fair Scheduler (PFS) also takes advantage of the changing radio conditions and channel diversity, but in addition to performing an efficient allocation of system resources, it also considers user fairness. This provides a middle ground between the RR and the Max C/I algorithms. The PFS algorithm assigns turn and codes to the different users based on a fair throughput and time approach as defined in the following equation:

$$P_i = \frac{IB_i[n]}{AB_i[n]} \tag{4.1}$$

Where
    P_i is the priority of user 'i' for the current transmission turn;
    IB_i is the instantaneous bit rate of user 'i'; and
    AB_i is the average throughput of the connection for user 'i'

IB_i can be calculated by the scheduler based on the CQI value reported by the UE, therefore better channel conditions lead to a bigger values of IB_i. However, this does not necessarily translate into a bigger AB_i value because the history of the UE's throughput is equally important in determining the user's priority in the queue.

As an example, consider that UE 'A' is a category 6 device (capable of 16QAM) and is in a good radio condition with a CQI value of 22. The scheduler finds at that particular location the UE can support up to 3 Mbps. Assume that the average throughput for this UE is 2.2 Mbps, in which case the scheduling priority value P_i is determined to be 3/2.2 = 1.36. Now consider a second user in the cell, UE 'B', who is a category 12 device (capable of QPSK) under poor radio conditions with CQI = 10. In that condition, the scheduler can only support up to 600 kbps for UE B; assuming that the average throughput of user B is 400 kbps, the scheduling priority is P_i = 600/400 = 1.5. In this case, the scheduler will select UE 'B' for fairness reasons, due to its higher priority P_i.

Figure 4.7 illustrates how the PFS algorithm works with two active users. The radio conditions vary for both User 1 and User 2, as depicted by the varying data rates that each user is capable of through time. The scheduled users in each time slot reflects the signal quality and fairness criterion noted.

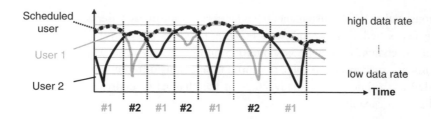

Figure 4.7   HSDPA Proportional Fair Scheduler

The efficiency of the scheduler algorithm depends on the cell load, traffic distribution (multiuser diversity) and the type of application being served; therefore, when comparing scheduler performance it is important to understand the conditions under which the comparisons are made. While theoretical studies suggest significant capacity gains for the PFS algorithm (on the order of 40–100%) [7] our tests show gains of up to 40%, but these are only possible in low mobility situations. When the mobiles move at medium speeds (50 km/h and above) the gain is reduced to about 5%. Chapter 6 provides more details on our lab and field tests comparing the algorithms.

4.2.2 HSUPA Scheduling

The HSUPA scheduler is completely different in nature from the one used for HSDPA, mainly because there is no time division multiplexing in uplink. Since all HSUPA users may be transmitting simultaneously through different channelization codes, the objective of the HSUPA scheduler is to distribute the common resource – power, in this case – among all the competing users. The implementation of the HSUPA scheduler is vendor dependent. As was the case with HSDPA, there are multiple techniques that can be applied, such as Round Robin (equal distribution of power among users), proportional fair [8] (power assignments based on average data rate and radio conditions) or more complex algorithms such as delay-sensitive schedulers that are appropriate for real-time services [9].

The HSUPA scheduler is located in the NodeB, which provides the advantage of a fast response to changing conditions. This scheduler can potentially operate on a TTI basis, which is 2 or 10ms, depending on the UE category. The scheduler decides how much power a UE can transmit. As with HSDPA, the scheduler's exact specifications are not defined in the standards; therefore the implementation is vendor dependent.

Figure 4.8 illustrates all the inputs that may be considered by the E-DCH scheduler. Among other information, such as each UE's radio conditions and the overall interference, the scheduler decisions may take into consideration the mobile requirement's to transmit at a higher bit rate. This is indicated to the NodeB through the 'happy bit' contained in the E-DPCCH channel (EDCH dedicated physical control channel). [10] The happy bit indicates whether the UE is satisfied with its current data rate, which is why it is called a happy bit.

The results of the scheduling decision can be communicated to the UE via two physical channels specified for HSUPA, the E-AGCH (Absolute Grant Channel) or the E-RGCH

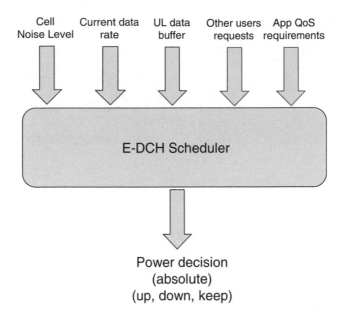

Figure 4.8 HSUPA scheduler inputs and outputs

(Relative Grant Channel). The Absolute Grant, controlled by the Serving E-DCH cell, provides the UE with an absolute value of the UL transmit power that can be used for data transmission. The Relative Grants are used by other cells in the active set, and provide relative increments or decrements over the current maximum UL power.

4.3 HSDPA Power Allocation

HSDPA users are allocated a share of the total power on the cell; power that is shared with voice users and Rel.'99 data traffic. As a rule of thumb, the higher the total power allocated for HSDPA, the higher the per user throughput and cell capacity.

The total power allocated for HSDPA is shared by all the users that have an active session in the cell, depending on the rules defined by the scheduler, as discussed in the previous section. There are two ways to allocate power to the HSDPA service, as illustrated in Figure 4.9:

- **Static Power Allocation (SPA):** the cell reserves a fixed amount of power for HSDPA out of the total available power amplifier power (for instance 6 watts); and
- **Dynamic Power Allocation (DPA):** the power used by HSDPA varies adapting to the available power amplifier power at every point in time.

In both cases, the real-time voice traffic from the dedicated channels will typically have priority over data traffic, so the power granted to HSDPA will be that remaining after the voice traffic has been allocated. In the existing implementations it is usual to define priorities between

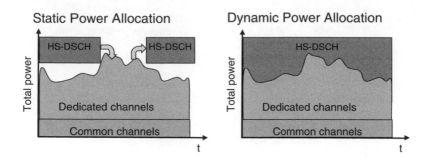

Figure 4.9   Illustration of Static vs. Dynamic Allocation

traffic types, which control the behavior of the power allocation strategy when distributing the power between Rel.'99 data users and HSDPA. Vendor specific implementations make it possible to define a minimum power to be allocated to HSDPA in order to support QoS guarantees for real-time services.

SPA is very simple to implement and control. Because the power for HSDPA is fixed, it is easy to predict the impact on the overall network performance and greatly simplifies the design. On the other hand, this is not very efficient, because it limits the maximum achievable HSDPA throughput. There are several possible variations of static power allocation, for instance reserved power for HSDPA can be de-allocated if the voice traffic increases beyond a certain point, or when there is no HSDPA traffic to transmit.

In the case of DPA, the power used by HSDPA varies with the time, depending on the resources allocated for the DCH traffic. This ensures an efficient utilization of the power of the cell, which will be dedicated to transmit a higher HSDPA throughput whenever there are resources available. For this same reason, DPA can also improve the coverage of the cell, because the users at the cell edge are not limited to a pre-defined amount of power for the traffic channel.

The fact that with DPA the cells will be transmitting at the maximum power can create a higher interference level in the network which can ultimately affect the quality and capacity for voice users. For this reason, vendors have incorporated measures to reduce the impact of the HSDPA traffic by either (a) limiting the maximum power for the HS-DSCH channel or (b) by implementing a power-control-like function in the HS-DSCH. Figure 4.10 demonstrates the latter algorithm (b), where the HSDPA users do not use all the remaining power. We call this the 'minimum power' strategy. In this case, the NodeB is intelligent enough to determine the amount of power needed by the active HSDPA users in the cell, and allocates the minimum resources required to achieve that goal.

The advantage of adopting a 'minimum power' strategy for HSDPA is an overall reduction of interference in the network, which reduces the impact on voice services. There is a risk, however, that such strategy can provide suboptimal HSDPA performance if the resources are not accurately calculated by the NodeB. Also, a highly varying HS-DSCH channel power may cause problems with the CQI estimation process, which can result in incorrect decisions for

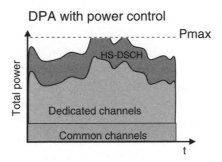

Figure 4.10 Illustration of Dynamic Power Allocation (DPA) with power control using the 'minimum power' strategy

TBF selection or scheduling. Real-life examples of these scenarios are provided through examples in Chapter 6.

4.4 Power Control and Link Adaptation

In order to achieve optimum performance in a wireless system, where the radio conditions are rapidly changing, methods must be implemented to compensate for these variations. Power Control (PC) and Link Adaptation (LA) are methods that help the connections overcome degraded channel conditions, or transmit at higher bit rates when the radio conditions are better. At the same time, PC and LA ensure that the overall consumption of valuable system resources is minimized.

The use of fast power control has been extremely successful in CDMA networks because it helps overcome fast fading and at the same time control the overall interference in the cell. Fast power control is used for voice and Rel.'99 data traffic as their main mechanism for control of the connection quality. HSUPA also uses fast power control together with other connection control mechanisms, such as LA. On the other hand, HSDPA connections do not use power control because the channel is shared among multiple users.

LA, also known as Adaptive Modulation and Coding (AMC) is a mechanism for the control of the connection quality that is especially suitable to packet switch data services. The AMC function adjusts to the changing radio conditions by increasing or decreasing the protection level of the transport blocks, through different modulation or redundancy levels. When channel conditions are ideal, AMC permits the user to transmit at the highest possible bit rate, and vice versa. This makes the most efficient use of the channel in the 'one big pipe' concept, as explained earlier.

In the following subsections explain the PC and LA mechanisms.

4.4.1 Power Control

As mentioned earlier, power control is not used in HSDPA due to the shared nature of the channel. However, it is used for HSUPA. Power control plays an important role in any CDMA system both at the connection level and at the cell level because it reduces the interference and hence minimizes the effective network loading. At the connection level, PC can overcome fading by rapidly adjusting the transmit power to match that needed to transport the data payload; on the cell level, it minimizes the overall interference level by reducing the transmit powers to the minimums necessary.

In HSUPA, the use of PC avoids the 'near-far' problem, in which a User Equipment (UE) far away from the NodeB cannot be heard by the NodeB's receiver because of the interference caused by UEs near the base station. To overcome this problem, the NodeB instructs those UEs that are close to the NodeB to reduce their transmit power while commanding those far away to power up. The idea is that the received power from all the UEs will arrive at approximately the same signal-to-interference ratios at the receiver front end.

There are two techniques to the power control mechanism: open loop and closed loop. Open loop is used during the initial access stage, while closed loop operates after the connection has been established.

Open loop is employed to determine the initial transmit power for a UE that wants to establish a new connection. During open loop power control, the UE estimates the required mobile transmit power by evaluating the received downlink signal. A UE needing to access the network will make one or more access attempts. Each access failure will cause the mobile device to power up a step, the size which is configurable by the operator. The maximum number of possible attempts is also a configurable parameter.

Closed loop power control becomes effective after the UE enters dedicated mode. It is a fast power control that operates at a frequency of 1500 Hz, meaning there are 1500 power updates per second. There are two layers in the closed loop operation: outer loop and inner loop, each one with a different target threshold. Because interference control is important in CDMA systems, the goal of the power control algorithm is to ensure that each user is served with the minimum power level required to achieve the target frame error rate (FER) or block error rate (BLER) which is directly related to the quality of the service. The target FER or BLER is configured by the operator for each type of service.

The BLER target for each type of service depends on the quality of service requirement, and in general is configurable by the operator. A tight BLER target can be used for services with stringent quality requirements, such as voice, while for services which can tolerate some data frame losses, such as non-real time data, the BLER target can be loose. The typical BLER target for voice is in the $1 \sim 2\%$ range; data services, on the other hand, can rely on retransmissions to recover the packet loss, and therefore their BLER target can be higher, with 10% being a typical value.

Fast power control is the foundation for traditional CDMA networks and it provides large voice capacity improvements; however, this method alone is not sufficient to ensure the quality of the connection for a high speed data service. The fundamental challenge is that for high speed data, the extreme bandwidth requirement effectively reduces the spreading gain provided by a CDMA system. Using a HSUPA bearer with SF 4 as an example, the processing gain is about 6 dB which is far less than that of a narrowband 12.2 k voice call with 21 dB processing gain. A reduction of the processing gain translates into a reduced link budget and higher power consumption per radio link for data services. Those fundament challenges have forced the industry to find new solutions to complement the next generation data services, such as link adaptation and packet scheduling.

4.4.2 Link Adaptation

LA is used by both HSDPA and HSUPA, where it provides the ability to adaptively use different modulation or channel coding schemes. These can be changed every couple of milliseconds, depending on the radio conditions. LA in HSPA replaces the previous, less efficient channel adaptation method based on modification of the spreading factor used by Rel.'99 data services.

In general terms, the LA algorithm will evaluate the current radio conditions, the amount of data to be transmitted by the user, and the existing resources to select the most appropriate Transport Format Block (TFB) to be transmitted at every moment, as shown in Figure 4.11. As the channel conditions change, the mix of channel protection using Forward Error Correction (FEC) and modulation versus the payload data in the Transport Block size is varied. In good conditions, as measured by a high Channel Quality Indicator (CQI), little channel protection is needed and larger block sizes can be used. However, in poor conditions, denoted by low CQI values, more channel protection is needed and transport block size correspondingly reduced. Throughput (instantaneous bit rate) goes down in the poor conditions, but the data continues to be transported with high quality due to the increased protection, resulting in a relatively constant BLER. As noted, the LA process in HSPA is very fast and updates can take place every Transmission Time Interval (TTI) (2 ms).

The efficiency of the LA function in HSPA is further enhanced through Hybrid Automatic Repeat-reQuest (HARQ) processes, which, as described in Chapter 2, provide a way to combine information from successive packet retransmissions, thus improving the decoding

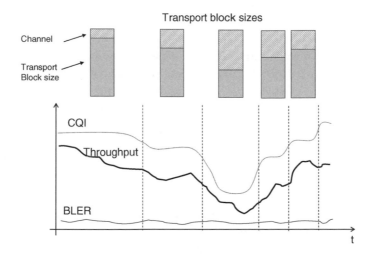

Figure 4.11   Example of Link adaptation for HSDPA using a single modulation scheme

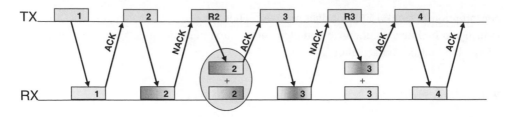

Figure 4.12 Illustration of HARQ functionality with acknowledgements (ACKs) and negative acknowledgements (NACKs) controlling retransmissions

success rate. Studies by 3GPP companies have estimated that the utilization of HARQ can provide more than 20% capacity improvement with HSDPA [2].

HARQ reduces the number of required modulation and coding scheme (MCS) levels and the sensitivity to measurement error and traffic fluctuations. HARQ is implemented on Layer 1 and resides in the NodeB, which greatly reduces the measurement reporting cycle and acknowledgment time. Figure 4.12 illustrates the concept of HARQ, where partially incorrect packets can be combined to obtain one that is free of errors; furthermore, with AMC the packets can be retransmitted with a different, more robust coding than the initial transmission, further increasing the probabilities of decoding with successive retransmissions.

4.4.2.1 Link Adaptation in HSDPA

In the case of HSDPA the LA function selects the proper Transmit Block (TB) sizes, coding rates and modulation schemes to match the channel conditions. Because the downlink shared channel (HS-DSCH) is a common channel to all active users, the downlink scheduler works together with the AMC function to ensure the system delivers service to all users quickly and fairly. The LA process assumes that the HSDPA channel power is either fixed or stable over a certain adaptation period.

HSDPA blocks have a large variation range, depending on the number of parallel codes (1–15), the modulation used (8PSK or 16QAM) and the amount of FEC redundancy. Although higher modulation schemes provide the potential of delivering higher throughput, these require better good channel conditions (CQI > 16) which can only be possible in certain parts of the cell coverage area. In a field environment, as we will present in later chapters when field results are discussed (see Chapter 6), the percentage of areas where the mobile device can use higher modulation schemes such as 16QAM is around 20–50%, although this also can increase with the use of advanced receivers.

The implementation of AMC for HSDPA requires new functions be added to the UEs to measure and report the channel quality and send them back to the NodeB. For this purpose, 3GPP introduced a new measurement unit called *Channel Quality Indicator* (CQI). The NodeB determines the data rate and TB size to be delivered based on the CQI value provided by the UE. Table 4.1 is an example of CQI values mapped into different TB sizes for UE categories 1 to 6 [3].

Table 4.1 CQI mapping table for UE categories 1 to 6

CQI value	Transport Block Size	Number of HS-PDSCH	Modulation	Reference power adjustment Δ
0	N/A		Out of range	
1	137	1	QPSK	0
2	173	1	QPSK	0
3	233	1	QPSK	0
4	317	1	QPSK	0
5	377	1	QPSK	0
6	461	1	QPSK	0
7	650	2	QPSK	0
8	792	2	QPSK	0
9	931	2	QPSK	0
10	1262	3	QPSK	0
11	1483	3	QPSK	0
12	1742	3	QPSK	0
13	2279	4	QPSK	0
14	2583	4	QPSK	0
15	3319	5	QPSK	0
16	3565	5	16-QAM	0
17	4189	5	16-QAM	0
18	4664	5	16-QAM	0
19	5287	5	16-QAM	0
20	5887	5	16-QAM	0
21	6554	5	16-QAM	0
22	7168	5	16-QAM	0
23	7168	5	16-QAM	−1
24	7168	5	16-QAM	−2
25	7168	5	16-QAM	−3
26	7168	5	16-QAM	−4
27	7168	5	16-QAM	−5
28	7168	5	16-QAM	−6
29	7168	5	16-QAM	−7
30	7168	5	16-QAM	−8

© 2008 3GPP

The AMC function takes the CQI reports and estimates the data rate that can be delivered under the conditions. Since the selection of modulation and coding schemes is based on the feedback received by the system, timely and accurate channel estimation is critical for the AMC function. Errors in the channel estimation can lead the AMC function to select the wrong modulation and coding rate combination, and the system will be either too aggressive or too conservative when assigning the transmit block size. This results in either capacity loss in the case of conservative decisions, or high BLER in the case of aggressive decisions. Furthermore, the 3GPP standard does not provide requirements on how to calculate the CQI value, which results in different UE devices exhibiting different performance under similar radio conditions.

In order to compensate for the inaccuracy of the channel measurements and the inconsistency among UEs vendors, LA algorithms usually offer a possibility for CQI adjustment. When this feature is activated, the RAN will try to maintain a preconfigured retransmission rate, for instance 15%, by adding an offset to the CQI value reported by the UE. If the retransmission rate is higher than the target, the CQI offset will be a negative value, which leads to the assignment of smaller TB size and better coding protection for the data being transferred. The target retransmission rate has a direct impact on the HSDPA throughput, but it is a hard coded value in most implementations and therefore operators cannot control it. Too aggressive retransmission targets will lead to higher packet losses and retransmission rates, while too conservative targets will underutilize system resources – both of which reduce the overall system capacity [4]. Chapter 6 presents results from different LA schemes that demonstrate the tradeoffs between target retransmission rate and system capacity.

4.4.2.2 Link Adaptation in HSUPA

In the case of HSUPA LA is performed through adaptive coding and HARQ combined with fast power control. The LA in HSUPA is conceptually the same as with HSDPA, with the following differences:

- In HSUPA the power of the link will vary with fast power control, while in HSDPA it was constant or very stable.
- There is no adaptive modulation in HSUPA since the Rel.'6 specifications only define one modulation type for uplink (BPSK). This will change in future releases, when additional modulation types are allowed (see Chapter 8).
- There is no need to send CQI quality measurements, because the NodeB is the element receiving the data and therefore the decision can be more accurate than HSDPA CQI measurements sent by the handset.
- HSUPA can operate in soft handover so the HARQ process may need to combine information from different cells.

As it was the case with HSDPA, the LA algorithm will select an appropriate transport block size based on the allowed uplink power for transmission. This information will be communicated by every TTI to the UE via the E-TFCI (Transport Format Combination Indicator) [5], which can indicate up to 128 different transport block formats. Table 4.2 provides the maximum amount of bits that can be transmitted by each HSUPA terminal category in one transport block [6].

One important aspect to consider is the tight interaction between the LA, the HSUPA scheduler and the fast power control mechanisms. These three mechanisms are working towards the same goal, effectively increasing uplink capacity and quality. The scheduler mechanism controls the power resources of all users in the cell, attempting to create a balance between user demands (data rate) and overall interference. The fast power control manages the power of a specific user, establishing tradeoffs between user demands (data rate) and channel

Table 4.2 HSUPA UE categories

E-DCH category	Maximum number of E-DCH codes transmitted	Minimum spreading factor	Maximum number of bits of an E-DCH transport block transmitted within a 10 ms E-DCH TTI	Maximum number of bits of an E-DCH transport block transmitted within a 2 ms E-DCH TTI
Cat 1	1	SF4	7110	(Only 10 ms TTI)
Cat 2	2	SF4	14484	2798
Cat 3	2	SF4	14484	(Only 10 ms TTI)
Cat 4	2	SF2	20000	5772
Cat 5	2	SF2	20000	(Only 10 ms TTI)
Cat 6	4	SF2	20000	11484

conditions. Finally, LA works essentially towards the same goal as the power control, finding the most appropriate coding for a given power and radio condition. In practice, these functions work at different stages of the transmission, as illustrated in Figure 4.13:

- Packet scheduler, typically working 2 or 10 ms TTI periods, decides the power to be used by the UE.
- Link adaptation, working at 2 ms TTI periods, decides the coding format based on the radio conditions and allocated power.
- Power control, working at 0.66 ms (1/1500 Hz), adapts to overcome fading and short-term interference.

The packet scheduler and LA functionalities are very much coupled, because the HSUPA scheduler has one feature that can limit the transport channel formats (TCF) to be used by the link adaptation.

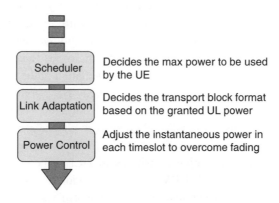

Figure 4.13 Interaction between Link Adaptation, Scheduler & Power Control

4.5 Mobility Management

Mobility management procedures control the connection during the cell transitions, ensuring that the data communications continue after a handover to a neighbor cell. The goal of mobility management is to ensure the handovers occur with minimum or no data loss, and minimum interruption delay. The improvement in data mobility performance is a significant differentiator between the GSM and UMTS data experiences. While in EDGE or GPRS, the cell transitions are performed through cell reselection procedures, which need to close the connection in the serving cell before establishing the call in the target cell. In contrast, UMTS Rel.'99 utilizes handovers (soft and hard) for data connections, which means that the target cell's resources are established prior to the handover and there is a minimal interruption during the transition.

While HSUPA mobility procedures are very much based on the Rel.'99 procedures, the shared nature of the HSDPA channel required a different mechanism that is called HS-DSCH cell change. Operators need to know that while the mobility procedures are defined in the standards, the vendors provide differing implementation algorithms. Depending on the implementation, there could be longer or shorter interruption times during cell changes. The following subsections explain both HSDPA and HSUPA mobility management procedures and the different implementation methods.

4.5.1 HSDPA Mobility Management

HSDPA channels are time and code shared between multiple users in each sector, making it very difficult to synchronize the allocation of users between different cells. This, together with the fact that the shared channel is not power-controlled, is the main reason why there is not a soft handover mechanism in the HS-DSCH channel. On the other hand, the associated control channel (A-DCH) that is always established during a HSDPA connection can support soft handover, which facilitates the exchange of control information during the cell transition.

Using hard handovers instead of soft handovers means that there will be some interruption in the data traffic flow at the transitions, which can potentially lead to data loss. If these transitions are not implemented in an efficient way, then the interruption delay can be noticeable for the user. In addition, the interruptions could prevent an operator from offering real-time services that have very stringent handover requirements. The HSDPA specifications have introduced mechanisms to permit an efficient transition while transmitting on the high-speed channel; however, much of the final performance depends on the vendor's implementation of the different options.

The cell transition mechanism in HSDPA is called 'HS Serving Cell Change'. The implementation of this procedure is optional, and the infrastructure vendor may opt to perform HSDPA cell changes via transitions to Rel.'99 dedicated channels. Such an arrangement is highly inefficient but may be acceptable for networks in the early deployment state. With an increased number of users performing transitions to DCHs, the result can be noticeably decreased performance during the HSDPA transitions. In addition a significant degradation in sector capacity occurs if those DCH transitions are frequent. These two options are illustrated in Figure 4.14.

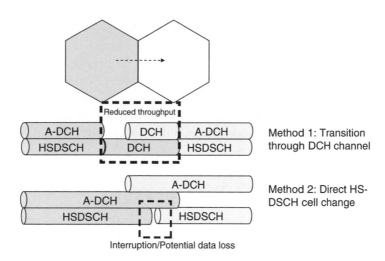

Figure 4.14   Cell transition mechanisms with HSDPA illustrating two different methods

In HSDPA the connection is established for only one cell at a time, which is called the 'Serving Cell'. With the first release of HSDPA (Rel.'5) the network decides at every time what is the appropriate serving cell for each UE, based on the measurements reported. Rel.'6 introduced the possibility to induce a serving cell change based on UE decisions.

When the network considers that there is a better serving cell than the current one, it sends an indication back to the UE to perform a HS-DSCH cell change. The actual cell change action may be synchronized, where the transition happens at a time indicated by the network, or unsynchronized, where the UE is asked to transition 'as soon as possible'. When the cell change action is triggered, the UE stops receiving to the current HS-DSCH channel and tunes to the one in the target cell. During this transition there may be data loss because some of the frames arriving to the previous serving cell will be left in the MAC-hs queue of the old NodeB and would not be delivered in the target cell. This would typically happen in the case of inter-NodeB cell changes; however, in the transition between cells belonging to the same NodeB (intra-NodeB cell changes) the MAC-hs data does not need to be flushed, so the data can be retained. See Chapter 2 for further details on MAC protocol architecture for HSDPA.

The transition between NodeB's belonging to different RNC present a more complex scenario in which the MAC-d layer need also be flushed and relocated to the new target RNC. In real deployments, this transition presents considerably more degraded performance than the intra-RNC ones. Different strategies may be adopted depending on the vendor as to when to 'handover' the control to the new RNC:

- **Fast Target RNC transition through SRNS relocation:** the HS-DSCH cell transition in the new RNC triggers a reassignment of the controlling RNC.
- **Delayed Target RNC transition with Iur mobility:** the 'old' RNC keeps control of the HSDPA MAC-d layer and routes the traffic through the Iur to the new RNC. The new RNC will typically take control when the HSDPA data buffers are empty.

The main advantage of the fast target RNC method is that the new RNC and Serving Cell have full control over the assignment of the resources related to the ongoing HSDPA connection. The main disadvantage is the added complexity, and potential delay introduced by the additional procedure. The main advantage of the delayed target RNC method is that while the old RNC does not have full control over the resources in the new cell, this method can potentially reduce the interruption during the cell transition because of the reduced complexity.

As it can be seen, the cell change mechanism in HSDPA is far from ideal and presents several challenges, including the following:

- There will typically be data loss or, in case the information is retransmitted, interruption times that can range from tens of milliseconds to a few seconds.
- The absence of soft handovers also presents a radio challenge because the interference created by the mobiles at the cell edge will be quite high until the transition is finalized. There will be abrupt changes in the radio condition that can have a significant effect on the performance, especially when the HSDPA cells are not loaded with data traffic and a HSDPA dynamic power allocation strategy is used.

On the other hand, the performance of HSDPA cell selections is likely to improve in the near future when new standard changes become commercially available. For instance, Rel.'6 introduces a method to bi-cast packets to both NodeBs involved in the procedure (support of out-of-sequence PDUs) which eliminates the data loss during the MAC buffer reshuffling.

4.5.2 HSUPA Mobility Management

The mobility management in HSUPA does not include any major modifications compared to the previous Rel.'99 DCH channels. With HSUPA soft-handovers on the traffic channel are still possible and several NodeBs can receive the uplink data, combining the different streams to achieve better decoding performance. Figure 4.15 illustrates this concept.

Soft-handover is particularly critical in the uplink to avoid the near-far problem, in which a mobile would approach a neighboring cell with a very high uplink power. This could subsequently increase the noise rise and create interference effects to the users in that cell. In HSUPA, the mobile can transmit to two or more simultaneous cells at the same time, and the cell with the best path loss determines the power transmitted by the mobile, according to the power control procedures discussed. Although the data is being received by different sectors, it can still be combined to achieve a reduced error rate: Maximal Ratio Combining (combination at the signal level) can be applied if the cells belong to the same NodeB, while Selective Combining (combination at the frame level, performed at the RNC) can be used for cells belonging to different NodeBs.

The cells to which the mobile can transmit are called the 'active set'. In HSUPA there can be up to four cells in the active set, while Rel.'99 allowed up to six. As in the case of Rel.'99 channels, cells enter and leave the active set through radio link additions, deletions and replacements. One of the cells from the active set will be the 'serving cell', which as explained in Section 4.4.2 is in charge of controlling the overall power allocation (data rate) for the user.

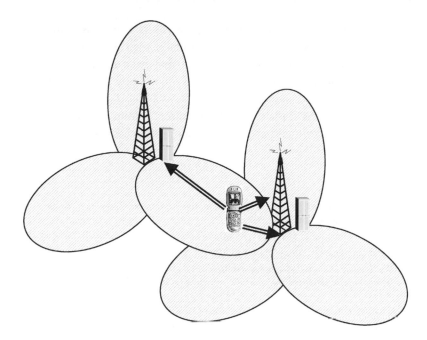

Figure 4.15   Illustration of soft-handover with HSUPA

Although not necessary, it is typical that the HSDPA and HSUPA serving cells are the same one, and therefore the change of Serving E-DCH cell will occur at the time of a HSDPA Serving Cell Change.

Soft-handover provides improved user performance thanks to the soft-combining gain. However, while soft handover is a good feature to have, allowing many cells in the active set has its drawbacks. Too much of a good thing is not that good after all, as there is payment in terms of additional baseband capacity being used in multiple cells at the same time, which is especially harmful during busy hours. This is one of the reasons why the Soft Handover Overhead should be carefully reviewed when HSUPA service is activated. On the other hand, there are alternatives to overcome the increased capacity demands from HSUPA soft handovers; some vendors allow the use of soft-handover on the control channel, while the traffic channel will effectively be processed only by one of the sectors, the Serving E-DCH cell. Such an arrangement is still compliant with the 3GPP specifications and can save a significant amount of baseband resources in the soft-handover areas.

4.6 Summary

The main takeaways and lessons learnt throughout this chapter were:

- In HSPA architecture, the RRM functionality has been moved to the NodeB level compared with Rel.'99. This enhancement enables faster responses to changing radio conditions and results in improved overall capacity and quality.

- HSPA provides a big pipe that operator should use wisely – not wastefully.
- HSDPA introduces new Radio Resource Algorithms such as Fast Packet Scheduling, HSDPA Power Allocation and Link Adaptation that have a significant effect on the performance – operators should carefully review these new items to understand their behavior.
- Mobility procedures in HSDPA and HSUPA are different from Rel.'99 channels and their limitations should be understood and considered during network planning and optimization.

References

[1] Holma, H. and Toskala, A., *WCDMA for UMTS. Radio Access for Third Generation Mobile Communications* (2nd Edition), John Wiley & Sons Ltd, 2002.

[2] 3GPP Technical Specification 25.848 'Physical Layer Aspects of UTRA High Speed Downlink Packet Access'.

[3] 3GPP Technical Specification 25.214 'Physical Layer Procedures (FDD)'.

[4] Tapia, P., Wellington, D., Jun, Liu., and Karimli, Y., 'Practical Considerations of HSDPA Performance', Vehicular Technology Conference, 2007. VTC-2007 Fall. 2007 IEEE 66th Sept. 30 2007–Oct. 3 2007. Page(s): 111–115.

[5] 3GPP Technical Specification 25.321 'Medium Access Control (MAC) protocol specification'.

[6] 3GPP Technical Specification 25.306 'UE Radio Access capabilities'.

[7] Kolding, T.E., 'Link and system performance aspects of proportional fair scheduling in WCDMA/HSDPA', Vehicular Technology Conference, 2003. VTC 2003-Fall. 2003 IEEE 58th Volume 3, Date: 6–9 Oct. 2003, Pages: 1717–1722 Vol. 3.

[8] Rosa, C., Outes, J., Dimou, K., Sorensen, T.B., Wigard, J., Frederiksen, F., and Mogensen, P.E., 'Performance of fast Node B scheduling and L1 HARQ schemes in WCDMA uplink packet access'; Vehicular Technology Conference, 2004. VTC 2004-Spring. 2004 IEEE 59th Volume 3, 17–19 May 2004 Page(s): 1635–1639 Vol. 3.

[9] You Jin Kang, Junsu Kim, Dan Keun Sung, Seunghyun Lee, 'Hybrid Scheduling Algorithm for Guaranteeing QoS of Real-Time Traffic in High Speed Uplink Packet Access (HSUPA)'; Personal, Indoor and Mobile Radio Communications, 2007. PIMRC 2007. IEEE 18th International Symposium on 3–7 Sept. 2007 Page(s): 1–5.

[10] 3GPP Technical Specification 25.212 'Multiplexing and Channel Coding'.

5

HSPA Radio Network Planning and Optimization

When operators plan to deploy HSDPA and HSUPA services, there are a number of tasks that need to be undertaken to achieve the optimum service performance in terms of quality, capacity and efficiency. These tasks include link budget analysis and cell network planning, followed by deployment and optimization of the network. During these deployment exercises, it is important to understand the challenges posed by the new HSPA technology. Being well prepared, knowing the right steps and understanding the tradeoffs will limit the amount of engineering effort required to optimize the network and ultimately reduce the overall deployment time and operation cost.

This chapter provides network operators with practical techniques for the typical planning and optimization tasks incurred with a HSPA network, highlighting the differences from legacy Rel.'99 UMTS practices. The proposals include the use of automated tools to extract maximum capabilities out of the technology. The chapter begins with a brief summary of the major differences between HSPA and Rel.'99 channels, introducing the main performance aspects where the operator should focus planning and optimization efforts. Next, in Section 5.2 details are provided for HSDPA and HSUPA link budget analyses, including several relevant examples. Section 5.3 presents a summary of simulation results for HSDPA and HSUPA, which helps readers obtain an understanding of the expected data capacity in the network. An overview of the network planning process is presented in Section 5.4. The remaining sections are focused on optimization of the network after the initial deployment, with Section 5.5 covering the optimization process through cluster drives, Section 5.6 including a discussion on the main parameters affecting the performance of HSPA traffic, and Section 5.7 presenting a methodology for automated parameter optimization which is very useful for networks that have reached mature levels of data traffic.

HSPA Performance and Evolution   Pablo Tapia, Jun Liu, Yasmin Karimli and Martin J. Feuerstein
© 2009 John Wiley & Sons Ltd.

5.1 Key Differences Between HSPA and Legacy Rel.'99 Channels

Prior to initiating planning for a UMTS network that will carry HSPA data traffic, it is important to understand the differences between HSPA and the legacy Rel.'99 channels, including the implications for the network planning and optimization. In many cases, UMTS networks were deployed before HSDPA or HSUPA were commercially available, thus the initial planning exercise was performed purely based on Rel.'99 services. When considering HSPA versus Rel.'99, not only are the radio channels different, but the traffic and usage characteristics of high speed data can also be vastly different.

When planning HSPA networks, or upgrading from a Rel.'99-only network, design engineers must pay careful attention to certain key differences from legacy UMTS networks, such as the different link level performance of the services, or the different mobility requirements. This last one, for example, would require smaller soft-handover areas as compared to typical Rel.'99 planning. Another example would be the fact that HSDPA makes use of Adaptive Modulation and Coding (AMC) schemes instead of power control. For this reason, no power control headroom needs to be considered in the link budget analysis for HSDPA traffic. In the following subsections we will discuss many of these differences and why it is important to consider them.

5.1.1 HSPA Data User Behavior Compared to Rel.'99 Voice Users

Typically UMTS networks have been planned and optimized for voice traffic. When the HSPA functionality is activated in the network, it is anticipated that the amount of data traffic will increase significantly. At some point data traffic may surpass that of voice, although this crossover point depends on the operator's data strategy.

Traffic characteristics are quite different between voice and data. Data typically is highly asymmetrical with more traffic on the downlink compared to the uplink. For this reason the deployment, performance and optimization of HSDPA will usually take precedence over HSUPA.

The bursty nature of data traffic together with the availability of higher data rates results in higher instantaneous transmit powers, which can raise the interference levels over short time periods. This can cause quality degradations to existing Rel.'99 voice users in the cell; therefore careful planning is required to balance voice and data performance and manage the associated tradeoffs.

5.1.2 HSPA Radio Performance Considerations Compared to Rel.'99

5.1.2.1 Spreading Factors

Both HSDPA and HSUPA use lower spreading factors compared to their Rel.'99 predecessors. This enables the much higher data rates available with HSPA, but entails inevitable tradeoffs. HSDPA uses a fixed Spreading Factor (SF) of 16, while HSUPA can use SF2 or SF4. In both cases, the smaller spreading factor implies a lower spread spectrum processing gain.

This processing gain deficit is compensated by the increased amount of transmit power allocated to the user with HSDPA. The lower processing gain is also mitigated by the improved MAC layer mechanisms in HSDPA, such as hybrid automatic repeat-request (HARQ), which makes the communication link more robust. Due to the use of HARQ, the HSPA data transmissions can operate at higher BLER levels compared to the legacy Rel.'99 channels to achieve a similar user experience. HSDPA can operate at BLER target levels of 15% or even 20% as compared to 10% for Rel.'99 channels.

5.1.2.2 Quality Metrics (Ec/No vs. CQI)

One important aspect regarding HSDPA performance analysis is the information provided by the Ec/No (energy-per-chip to noise spectral density ratio) metric, which is a measure of signal-to-noise-plus-interference ratio. The common pilot channel Ec/No is the key capacity and quality indicator in a UMTS network because it provides information regarding the relative increase of interference – and hence traffic loading – in the network. However, when analyzing HSDPA drive test data, common pilot Ec/No measurements must be carefully interpreted because they can provide a false indication of network quality.

In order to obtain an accurate estimate of the network quality, it is important to take into account the amount of transmit power from the sector that has been devoted to HSDPA during the tests. For example, consider an isolated cell with 10% of the cell power allocated to the pilot channel. When there is HSDPA traffic in that cell, the sector will allocate all the available power for HSDPA (assuming Dynamic Power Allocation (DPA) is activated) and the total transmit power will effectively reach 100%, in which case the pilot Ec/No measured by the UE will be at most $-10\,dB$. This occurs due to the low ratio of the sector's pilot-to-total transmit power (10% for the pilot channel $= 1/10 = -10\,dB$ pilot Ec/No without interference from other cells). When interpreting the measurements collected in this test, the pilot Ec/No values collected during the HSDPA transmission could mask the true radio conditions for the cell. In order to gain a clearer understanding of the radio conditions, the engineer should use the pilot Ec/No measurements collected in idle mode, when the HSDPA has not been activated. Idle mode measurements reflect the pilot Ec/No degradations due to coverage and interference effects, without the effects of loading from HSDPA.

Alternatively, the Channel Quality Indicator (CQI) provided by HSDPA can be a useful measurement to provide information regarding the radio conditions in the cell. The CQI value should follow the Signal-to-Interference plus noise ratio (SINR) measured by the UE, and as such is an important metric to consider during the network optimization. In commercial practice, interpreting the CQI reports shouldn't be taken as the unique metric for quality due to the loose manner in which it has been standardized. Different manufacturers and models of handsets may report different CQI values in the same radio conditions. We recommend that CQI measurements are benchmarked with idle mode Ec/No measurements to estimate the error deviation, if any.

5.1.3 HSPA Mobility Considerations Compared to Rel.'99

The mobility procedures for HSDPA and HSUPA are different from those of Rel.'99 channels. In general, soft-handover in HSDPA and HSUPA present more challenges than benefits. For HSUPA, soft-handover represents a potential network element capacity problem because it can result in excessive baseband hardware consumption for all the cells in the active set. In the case of HSDPA, cell transitions are not perfect and even the soft-handover addition and deletion of branches to the A-DCH active set can lead to degraded performance. The potential of constant ping-pong triggering of cell changes presents another challenge for HSDPA optimization in the soft-handover zone. In general, it is not desirable to have too much cell overlap, and hence large soft-handover regions, for cells supporting HSPA.

Inter-RNC HSPA transitions should also be minimized, at least within the core traffic areas. As explained in Chapter 4, inter-RNC mobility mechanisms imply higher complexity than the regular Serving Cell Changes. Upcoming material in Chapter 6 shows that this will typically result in reduced throughput or long (e.g., several seconds) interruptions at the RNC borders.

Finally, it should be considered that most HSDPA networks do not support direct transition to (E)GPRS; therefore the Inter-Radio Access Technology (Inter-RAT) boundaries need to be identified and carefully planned. As shown in Figure 5.1, we recommend that a Rel.'99-only buffer zone be established at the border between 3G and 2G to allow for seamless transitions from HSPA to Rel.'99 to (E)GPRS. At later stages HSPA can be activated in the buffer areas when the HSDPA compressed mode feature becomes available (expected around the 2010 time frame), or after appropriate tuning of the HSPA to Rel.'99 transition parameters.

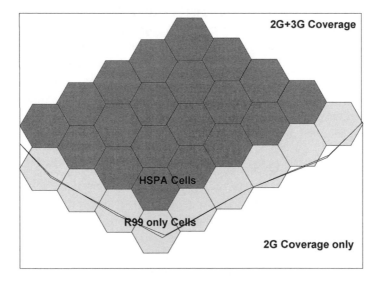

Figure 5.1 Illustration of buffer area of Rel.'99 at the edge of 3G coverage between HSPA and 2G (E)GPRS to facilitate seamless transitions

5.1.4 HSPA Baseband and Backhaul Resource Considerations Compared to Rel.'99

From a practical point of view it is important to consider that HSPA typically has a different utilization scheme of the baseband resources than Rel.'99 services. When provisioning for HSPA a significant amount of baseband resources will have to be devoted to this service, even if the HSPA traffic is minimal. This occurs because a dedicated packet scheduler is required in each NodeB (or even worse, in some implementations at the sector level). Of course, as the traffic demand grows more baseband resources will be necessary.

Similarly, the provisioning of backhaul capacity is a critical consideration that needs to be carefully planned. With the high data rates achievable by HSPA it is easy to create a bottleneck on the Iub backhaul interface between the NodeBs and RNCs. This is especially true for operators who use telco leased lines instead of owning their own backhaul transport infrastructure. As in the previous case, the backhaul provisioning should establish a tradeoff between cost, sector capacity and offered user data rates. In any case, backhaul deployment guidelines need to be considered during the RF planning phase. Chapter 7 includes a section devoted to capacity planning that covers both of these aspects.

5.2 Link Budget Analysis

As with any other wireless air interface deployment, link budget analysis should be the starting point for HSPA planning. The result from a link budget analysis allows operators to have high level estimates of the cell ranges for different data services. This helps operators select the best locations for the base stations. For operators who plan to overlay HSPA on top of an existing network (GSM or UMTS), the link budget analysis allows them to create a comparison of the coverage footprints between HSPA and the underlay network.

An analytical link budget calculation based on assumptions provided by the 3GPP standards or simulation results is always a good starting point. However, we recommend that those assumptions, such as Eb/No target, handover overhead, etc., be verified with drive test field measurements collected during the early stages of deployment or from trial networks.

The objective of this section is to describe the specific details for link budget analysis with HSDPA and HSUPA. A more detailed explanation and analysis on link budget specific parameters for UMTS can be found in general radio planning books, such as [1,2]. In most modern networks, the final design is typically performed with the help of computer-assisted design tools, such as propagation analysis, Automatic Cell Planning (ACP) and network simulators. The use of these tools will be reviewed further in Section 5.4.

5.2.1 Link Budget Methodology

The objective of the link budget analysis is to determine the maximum allowable pathloss between the mobile and the base station for the required service level. Pathloss calculations will determine the maximum cell radius (i.e., cell coverage area) by applying a propagation model

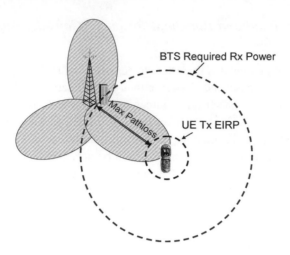

Figure 5.2 Illustration of maximum uplink pathloss determined by the UE maximum transmit EIRP and the base station required receive power

appropriate for the environment (urban, suburban, rural, etc.). In this section we will use an HSUPA uplink link budget calculation as an example to illustrate the general methodology.

In radio links, a mobile needs to reach the serving base station with a target required power level. Figure 5.2 shows these powers graphically in terms of Effective Isotropic Radiated Power (EIRP), which include the gains from the transmit antenna, the transmit-antenna cable and other losses. In the figure, the UE operates with a maximum transmit EIRP, while the base transceiver station (BTS) has a minimum require receive power level. The maximum pathloss is determined from these two contributing factors.

The link budget calculations begin with the service requirements at the base station, in terms of bitrate and required energy-per-bit to noise spectral density ratio (Eb/Nt). This Eb/Nt is subsequently translated into the required base station sensitivity, which is the required power at the receiver input. This can be translated into the minimum received power needed at the base station by taking into account the antenna gains, cable losses and other losses. Figure 5.3 illustrates the necessary steps to obtain this required minimum base station receive power value:

The BTS sensitivity can be calculated as:

$$\text{Receiver Sensitivity(dBm)} = \left(I + \frac{Eb}{Nt} - P_G \right)$$

Where

P_G is the processing gain $= 10 \cdot \log_{10}\left(\frac{W}{R}\right)$

I is the total UL interference(dBm) $=$ Thermal Noise $+$ Noise Rise $+$ K

K is the Noise Rise Adjustment Factor

Noise Rise $= 10 \cdot \log_{10}(1 - \text{UL Load})$

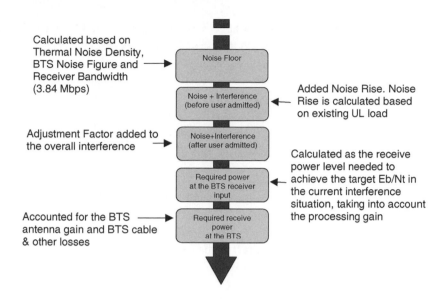

Figure 5.3 Calculation of required minimum receive power at base station

The EIRP values can be calculated as:

$$\text{BTS Required Rx Power} = (\text{BTS Sensitivity}) + (\text{BTS Antenna Gain})$$
$$- (\text{BTS cable \& other losses})$$

$$\text{UE EIRP} = (\text{UE Max Tx Power}) + (\text{UE Antenna Gain}) - (\text{UE cable \& other losses})$$

At the mobile, the antenna gain plus the cable, combiner and connector losses are typically close to a net value of 0 dB; therefore the EIRP will be equal to the maximum power. Finally, the total pathloss is calculated as the difference between the mobile EIRP and the BTS required receive power, plus certain design margins to account for slow fading variations and soft-handover gain [2]. These margins are calculated based on a desired cell-edge coverage confidence.

A similar approach can be applied for the downlink link budget calculation with some modifications. For instance, since there is no soft-handover in HSDPA, so handover margin should not be included in the calculation. Also, the maximum power for HS-DSCH can be much higher than that of the UE and power control is not used. There are some other major differences and we will go through all these details in Section 5.2.

5.2.2 Downlink Analysis

As noted, the link budget analysis for HSDPA is different from those performed for HSUPA or the Rel.'99 channels. This is because the power allocated to HSDPA will be the one determining the overall service experience. However, the selection of the HSDPA power can be based on two different planning targets: (a) the loading effect on existing voice services; or (b) the desired HSDPA bitrate at the cell edge.

Another difference is the metric used for target service quality. While Eb/Nt has been used previously as the design target, with HSDPA the use of Eb/Nt is not practical because this may change every TTI based on the modulation and coding used for different users. For this reason, HSDPA planning is rather based on the more general metric, Signal-to-Interference-and-Noise-Ratio (SINR), which is defined as [7]:

$$SINR = SF_{16} \times \frac{p_{\text{HSDPA}}}{(1-\alpha)P_{\text{own}} + P_{\text{other}} + P_{\text{noise}}} = SF_{16} \times \frac{p_{\text{HSDPA}}}{P_{\text{TOT\_BS}}(1-\alpha+\frac{1}{G})}$$

Where:

α is the orthogonality factor, which is 1 in the case of no-multipath and 0.5 for typical, non-rural multipath environments;
P_{own} is the power from the serving cell;
P_{other} is the interference power from the rest of the cells;
P_{noise} is the thermal noise power, N_o;
G is the Geometry Factor

The Geometry Factor (G) is one of the most important parameters affecting HSDPA throughput. G is defined as:

$$G = \frac{P_{\text{own}}}{P_{\text{other}} + P_{\text{noise}}}$$

The Geometry is a key aspect to consider in CDMA network planning. Its typical range is from −3 dB to 20 dB, depending upon the cell location and network geometry. More details on the nature of this factor can be found in [5] and its effect on the capacity of a WCDMA network can be read on [6].

Table 5.1 provides an indication of the expected HSDPA throughput depending on the SINR value, assuming HARQ and link adaptation [7].

Note that although the maximum achievable throughput would be around 11 Mbps (more than what is shown in Table 5.1), that would require a SINR of 40 dB or higher, which is not practical in real networks. The fact that the mobiles can consume multiple codes at the same time also needs to be considered when planning the network, since a proper balance

Table 5.1   HSDPA throughput vs. SINR (for 10% BLER)

SINR	Throughput with 5 codes	Throughput with 10 codes	Throughput with 15 codes
0	0.2	0.2	0.2
5	0.5	0.5	0.5
10	1	1.2	1.2
15	1.8	2.8	3
20	2.8	4.2	5.5
25	3.1	5.8	7.8
30	3.3	6.7	9.2

between Rel.'99 and HSDPA codes has to be established. If all 15 codes are reserved for HSDPA there will not be sufficient capacity left for voice calls in that sector.

5.2.2.1 Downlink Link Budget Example

In this section, an example link budget for a typical HSDPA deployment is presented. In this particular case, the maximum cell range is analyzed for the HSDPA service when the allocated powers for HSDPA are 6, 8 and 10 Watts out of a total of 20 Watts. The overall target downlink cell load (Common Control Channels + Rel.'99 + HSDPA traffic) is 90%. Considering that 25% power (5 Watts) is allocated to the pilot and other control channels, the corresponding Rel.'99 power assignments are 7, 5 and 3 Watts, respectively.

An orthogonality factor of 0.5 is assumed, with −3 dB of Geometry Factor corresponding to the edge of the cell under interference limited conditions. All this data can be used to calculate the HSDPA SINR at the cell edge. Once the SINR is known it is possible to estimate the maximum achievable bitrate at the cell edge from the information in Table 5.2.

Table 5.3 presents the final link budget calculations. With those calculations it will be possible to determine the radius for the cell edge.

Note that the maximum allowable pathloss in every case is 140 dB for different maximum bit rates, but for different transmit powers. This means the HSDPA bitrate at that pathloss varies depending on the allocated HSDPA power.

Another way to perform the link budget analysis is to estimate the bitrate that can be obtained at different pathloss values for a given HSDPA power. The following tables (Table 5.4, Table 5.5 and Table 5.6) illustrate the achievable bitrates for different use cases, given a maximum HSDPA power of 6 Watts for different geometry factors.

The results highlight the importance of the Geometry Factor in the achievable HSDPA data rates. If the cell edge is determined by the lack of UMTS coverage the achievable throughputs are significantly higher than in the case of transition to a new cell. This is one of the reasons to avoid too much cell coverage overlap with HSDPA.

5.2.3 Uplink Link Budget Analysis

In Section 5.2, we discussed the general methodology for uplink link budget calculation for a general UMTS system. The link budget analysis for HSUPA can apply the same methodology with certain exceptions. The two major differences to consider are the following: (1) the

Table 5.2 Expected HSDPA throughputs at the cell edge for different power allocations

HSDPA Power	SINR at cell edge	HSDPA throughput at cell edge
6 Watts	3.3 dB	330 kbps
8 Watts	4.5 dB	450 kbps
10 Watts	5.5 dB	550 kbps

Table 5.3 Example HSDPA link budgets for different bitrate requirements

	Parameters and calculations	Maximum cell edge bitrate		
		330 kbps	450 kbps	550 kbps
BTS Tx EIRP	Max NodeB Power (W)	20.0	20.0	20.0
	Max HSDPA Power (W)	6.0	8.0	10.0
	BTS Transmit Antenna gain(dBi)	18.0	18.0	18.0
	BTS cable connector combiner losses(dB)	4.0	4.0	4.0
	BTS EIRP(dBm)	51.8	53.0	54.0
UE Sensitivity	Thermal noise density (dBm/Hz)	−174.0	−174.0	−174.0
	UE Receiver Noise Figure(dB)	8.0	8.0	8.0
	Thermal noise Floor(dBm)	−100.2	−100.2	−100.2
	DL Target Load (%)	0.9	0.9	0.9
	Target Rise over Thermal (dB)	10.0	10.0	10.0
	Interference Floor (dBm)	−90.2	−90.2	−90.2
	Processing Gain (dB)	12.0	12.0	12.0
	Orthogonality Factor	0.5	0.5	0.5
	Geometry Factor	0.5	0.5	0.5
	SINR	3.3	4.5	5.5
	UE Rx Sensitivity(dBm)	−98.9	−97.7	−96.7
Maximum Pathloss	UE antenna gain(dBi)	0.0	0.0	0.0
	UE Cable connector combiner losses(dB)	0.0	0.0	0.0
	Slow Fading margin(dB)	−9.0	−9.0	−9.0
	Handover gain(dB)	0.0	0.0	0.0
	UE Body Loss(dB)	2.0	2.0	2.0
	Maximum Allowable Pathloss(dB)	*139.7*	*139.7*	*139.7*

existence of a power reduction factor or back-off with HSUPA; and (2) the increased overall interference level when a high-bitrate HSUPA user is admitted in the cell. If these factors are not considered, the link budget will provide optimistic results.

The high Peak-to-Average-Power-Ratio (PAPR) resulting from the use of multiple parallel codes requires the implementation of a power back-off in HSUPA that will keep the power amplifier operating in the linear zone. This back-off can range between 0 to 2.5 dB. 3GPP

Table 5.4 Bitrate achieved at different pathloss values for isolated cells (geometry factor, G, between 5 dB and 25 dB)

Pathloss	130	140	150
UE HSDPA Rx (dBm)	−80.2	−90.2	−100.2
UE Own Cell Interference (dBm)	−75.4	−85.4	−95.4
UE Other Cell Interference (dBm)	n/a	n/a	n/a
UE Noise (dBm)	−100.2	−100.2	−100.2
G (dB)	24.7	14.7	4.7
SNR (dB)	10.2	10.0	8.0
Bitrate at cell edge (kbps)	*1000*	*1000*	*800*

Table 5.5 Bitrate achieved at different pathloss values, for locations where two cells are received with the same signal strength (geometry factor, G, factor around 0 dB)

Pathloss	130	140	150
UE HSDPA Rx (dBm)	−80.2	−90.2	−100.2
UE Own Cell Interference (dBm)	−75.4	−85.4	−95.4
UE Other Cell Interference (dBm)	−75.4	−85.4	−95.4
UE Noise (dBm)	−100.2	−100.2	−100.2
G (dB)	0.0	0.0	−1.3
SNR (dB)	5.5	5.4	4.6
Bitrate at cell edge (kbps)	*550*	*540*	*460*

Table 5.6 Bitrate achieved at different pathloss values, for locations where three cells are received with the same signal strength (geometry factor, G, factor around −3 dB)

Pathloss (dB)	130	140	150
UE HSDPA Rx (dBm)	−80.2	−90.2	−100.2
UE Own Cell Interference (dBm)	−75.4	−85.4	−95.4
UE Other Cell Interference (dBm)	−72.4	−82.4	−92.4
UE Noise (dBm)	−100.2	−100.2	−100.2
G (dB)	−3.0	−3.1	−3.7
SNR (dB)	3.3	3.2	2.7
Bitrate at cell edge (kbps)	*330*	*320*	*270*

specifications [3] provide a method to calculate the back-off factor based on the channel configuration and desired bitrate. However, analysis on commercial HSUPA devices indicates that such method results in optimistic back-off calculations of around 0.6 dB [4]. As a quick reference of recommended practical values for back-off, Table 5.7 provides more realistic back-off values for several HSUPA target bitrates.

Traditionally, UMTS link budget analysis assumed that the noise rise in the system would not be significantly affected by the new user, which simplifies the overall interference calculations. While such an approach is valid for low bitrate applications, we have measurements that confirm that even a single HSUPA user can create a considerable increase in the uplink noise

Table 5.7 Eb/No vs. Throughput for a Category 5 HSUPA device (10 ms TTI, 1.92 Mbps Max Bitrate) [4]

Throughput (kbps)	Transport block size (bits)	Ec/Nt (dB) of 1st transmission	Required Eb/Nt (dB)	Commercial PA back-off (dB)
61	1026	−13.9	1.6	2.4
121	2052	−11.1	1.1	2.1
216	4068	−8.7	1.0	1.9
526	10152	−4.6	1.1	1.1
717	14202	−2.9	1.4	0.9
808	16164	−2.1	1.7	0.8

rise because high bitrates result in increased uplink power transmissions (see Chapter 6 for more details). To overcome this shortcoming, an Noise Rise (NR) adjustment factor can be added to the link budget calculation that effectively increases the overall noise rise according to the expected user data rate. This adjustment factor (K) is defined in [4] as:

$$\text{NRAdjustment Factor } (K) = 10 \cdot \log_{10}\left(1 + v \cdot \left(\frac{Eb}{Nt}\right) \cdot \left(\frac{R}{W}\right)\right)$$

Where v is the activity factor of the UL service, Eb/Nt[1] is the target service quality, and R/W is the processing gain. For simplicity purposes, in the rest of this section the application activity factor will be set to 1, a typical value for File Transfer Protocol (FTP) upload services. This corresponds to a conservative scenario compared to typical user scenarios.

The target service quality (Eb/Nt) is an important input for the link budget calculation, and is typically obtained from link level simulations. Table 5.7 provides suggested values of Eb/Nt for different target bitrates. The values provided are for a terminal with HSUPA Category 5 capabilities: 10 ms Time Transmit Interval (TTI), Spreading Factor 2, and maximum of 2 parallel codes.

The results in Table 5.7 are provided for a maximum of 2 HARQ retransmissions. With a maximum of 3 HARQ retransmissions the target Eb/Nt can be further relaxed. While a higher number of retransmissions can increase the cell range and the overall spectral efficiency, it could degrade the data experience because of the end-to-end delay increase caused by the higher percentage of retransmissions. This trade-off is reviewed in Section 5.6.

5.2.3.1 Uplink Link Budget Example

In this section we present an example link budget calculation for three different HSUPA data rates: 1 Mbps, 500 kbps and 64 kbps. The key factors used to compute the link budget are showed in Table 5.8. The target Eb/Nt values for these data rates are approximately 2 dB, 1.1 dB and 1.5 dB respectively, according to Table 5.7. From the same table, the corresponding mobile power back-offs are 0.6 dB, 1.1 dB and 2.3 dB, respectively.

The total cell load is designed to be 60% in the uplink, corresponding to a noise rise of 4 dB, which represents a conservative design approach. Based on our experience, typical noise rise values for mixed voice and data carriers can range between 4 and 6 dB, without overload effects. Data-only carriers with HSUPA could be configured to allow a 10 dB noise rise to maximize sector throughput.

Finally, the slow fading margin and handover margin are computed based on a 90% cell edge confidence, assuming a fading standard deviation of 7 dB, which is a typical value for an outdoor environment. These result in a −9 dB interference margin and 3.7 dB of macro diversity (handover) gain (for details on how to calculate these values see [2]).

---

[1] $Eb/Nt = Eb/(No + Io)$.

Table 5.8 Example link budget calculations for different uplink bitrates

	Parameters & Calculations	Target UL Data rates (kbps)		
		1000	500	64
UE Tx Power	UE Maximum Transmit Power (dBm)	21.0	21.0	21.0
	UE cable & other losses (dB)	0.0	0.0	0.0
	UE Transmit Antenna gain (dBi)	0.0	0.0	0.0
	Power back off factor (dB)	0.6	1.1	2.3
	Mobile EIRP(dBm)	20.4	19.9	18.7
Required Power	Thermal noise density (dBm/Hz)	−174.0	−174.0	−174.0
at the BTS	BTS Receiver Noise Figure(dB)	3.0	3.0	3.0
	Thermal noise Floor(dBm)	−105.2	−105.2	−105.2
	UL Target Loading (%)	0.6	0.6	0.6
	UL Noise Rise (dB)	4.0	4.0	4.0
	Required Eb/Nt (dB)	2.0	1.1	1.5
	Processing Gain (dB)	5.9	8.9	17.8
	Interference Adjustment Factor (dB)	2.5	1.2	0.2
	Node B Rx Sensitivity (dBm)	−99.8	−104.9	−113.9
Maximum Pathloss	BTS antenna gain (dBi)	18.0	18.0	18.0
	BTS Cable connector combiner losses (dB)	4.0	4.0	4.0
	Slow Fading margin (dB)	−9.0	−9.0	−9.0
	Handover gain (dB)	3.7	3.7	3.7
	BTS Body Loss (dB)	0.0	0.0	0.0
	Maximum Allowable Pathloss (dB)	*132.7*	*136.9*	*144.7*

As an example, assuming a commonly-used Hata slope-intercept model for pathloss, the results from this link budget exercise indicate that in a typical suburban macro cell environment it should be possible to achieve 1 Mbps bitrate (132.7 dB maximum allowable pathloss from Table 5.8) within distances beyond two miles from the serving cell. These results are aligned with our trial results on a HSUPA network with similar configuration, as shown in Figure 5.4 (see Chapter 6).

While this exercise is valid for outdoor locations, it is not designed for indoor scenarios. For indoor cases, the pathloss will increase due to building penetration losses (typically on the order

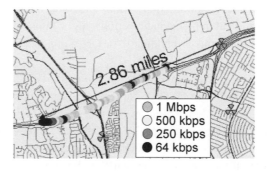

Figure 5.4 HSUPA field trial result in Suburban environment (Cat 5 UE)

of 15–20 dB) and the increased shadow fading margin. The typical standard deviation of the slow fading for indoor cases is close to 12 dB, which translates into a 15 dB fading margin for 90% cell edge confidence.

It is also worthwhile mentioning that due to the power back-off on the lower bitrates, the coverage with HSUPA 64 kbps is less than that achieved with the legacy Rel.'99 channels. Obviously, the advantage of HSUPA is the higher data rate capabilities.

5.3 Overview of System Level Simulations

Before any new technology is deployed or advanced features are standardized, infrastructure vendors and other R & D industry parties provide extensive simulation work to estimate the capacity and quality of the system under study. In the case of HSDPA/HSUPA, the most reliable information typically comes from studies based on dynamic system-level simulations, which incorporate many of the effects present in variable wireless environments, such as slow and fast fading, realistic behavior of radio protocols and algorithms, and to some extent realistic application models such as FTP, streaming, VoIP, etc. Unlike dynamic simulators, static simulations based on Monte Carlo analysis would fail to properly model features like fast fading, fast power control and link adaptation, fast NodeB packet schedulers, etc.

This section's review of existing simulation work provides a good estimate of the capacity of the network under loaded conditions, something impossible to infer from the simpler link budget analysis and difficult to accurately estimate based on purely analytical calculations. Having a sense of the overall bitrates the sectors can deliver and the expected per-user bitrates under loaded conditions are both critical metrics to understand before undertaking a network planning exercise. The simulation results also can be used as references to compare with the measurements from a network and give an indication on how well the network has been planned and optimized – how close the network is operating to the ideal performance.

When analyzing network simulation results it is important to clearly understand the conditions under which the results were modeled, such as assumptions about the following: coverage limited or interference limited scenario, data-only or mixed voice and data, one single data type or multiple services, QoS guarantees or best-effort, etc. Because simulation results are based on assumptions which do not necessarily fully reflect the actual network conditions, they should be interpreted as approximations. Many practical factors such as hardware and backhaul limitations, non-ideal implementations of the algorithms, rapid changes in propagation conditions, non-homogeneous in-building losses, etc. in general are not captured by the simulation tools.

Table 5.9 summarizes the effective HSDPA sector capacity under interference-limited conditions, from [8–10] regarding HSDPA capacity. Observe that the sector HSPA data capacities range from 1 to 2.2 Mbps based on the mix of voice and data loading and QoS parameters. As expected, the maximum capacity (2.2 Mbps) is achieved for the case where no QoS is guaranteed (i.e. best effort) and the HSPA carrier is dedicated to data-only traffic. As will be discussed in Chapter 6, these results are very much in line with the measurements gathered from our field tests.

Table 5.9 Expected HSDPA sector capacity for different data code usage and voice call scenarios based on simulations (from [8–10])

HSDPA (5 codes)	HSDPA (15 codes with best effort)	HSDPA (15 codes with 128 kbps guaranteed bitrate)	HSDPA (5 codes with 20 voice calls)	HSDPA (12 codes with 20 voice calls)
1.2 Mbps	2.2 Mbps	1.4 Mbps	1 Mbps	1.6 Mbps

Figure 5.5, taken from [10], provides an excellent illustration of the tradeoffs between HSDPA capacity and voice traffic in a sector. The figure plots HSDPA throughput versus Rel.'99 throughput. The cell can offer good HSDPA throughputs, of above 1 Mbps, even in relatively high voice load conditions (up to 30–40 Erlangs/sector, which is roughly equivalent to a Rel.'99 DCH cell throughput of 500 kbps).

In addition to gaining an understanding on the expected cell capacities, it is important to understand how the individual user bitrates respond to the overall data traffic load. The simulation study in [9] indicates that the median HSDPA bitrates reduce to half when the number of users in the cell increases from one to five. This non-linear effect is based on the multiuser diversity gain of the system, which leverages different radio conditions and traffic characteristics from different users transmitting simultaneously in the sector.

Turning to HSUPA, Table 5.10 summarizes the expected HSUPA sector capacity for different parameter settings [11,12] with mixed voice and data [13]. The simulation results in [14] indicate a total sector capacity between 1.2 Mbps and 1.8 Mbps, depending on the required latency target (affected by the number of retransmissions) for a 2 ms TTI network. The capacity of the system depends on the target BLER, number of retransmissions, and TTI length.

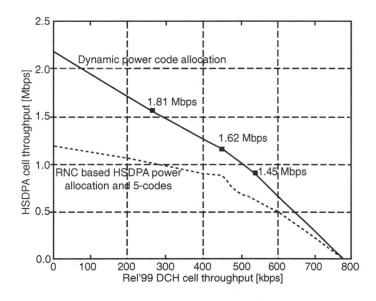

Figure 5.5 HSDPA cell throughput vs. Rel'99 traffic load (from [10]) © 2006 IEEE

Table 5.10 Expected HSUPA sector capacity with different parameter configurations for retransmissions and TTI times (Rtx = number of retransmissions) (from [11–13])

Data only 2 Rtx 10% BLER 10 ms TTI	Data only 0 Rtx 10% BLER 2 ms TTI	Data only 2 Rtx 10% BLER	Data plus 20 voice users 2 Rtx 10% BLER 2 ms TTI
1.5 Mbps	1.3 Mbps	1.8 Mbps	1.5 Mbps

While HSUPA sector capacity in a dedicated data-only carrier is less than that of an HSDPA data-only carrier, the gap closes when voice users are added within the same sector for mixed voice and data usage scenario.

5.4 Cell Planning Process

The cell planning process for CDMA technologies, such as UMTS/HSDPA, is more complicated than that with TDMA networks such as GSM. In UMTS, capacity, coverage and quality are not separable. Increases in amount of traffic carried by the network and the locations of the served users can cause shrinkage of the cell boundaries (called cell breathing). In addition, within the limits established by the baseband configuration and code resources, there is not a hard capacity limit for a UMTS cell. For instance, the number of voice calls which can be served by a UMTS cell is much higher when users are in good radio conditions, more than 60 users as we observed in our field tests, but this capacity decreases markedly if the users are located at the edge of the coverage area. Therefore CDMA networks have soft capacity limits.

The intrinsic relationship between traffic and transmit power usage in the cell will typically result in a frequent modification of the RF planning configuration to account for loading variations (e.g. interference levels, cell breathing). This is a completely different concept compared to other air interfaces, such as TDMA or even emerging OFDMA, where the RF planning is typically performed only once. With those air interfaces, increases in loading for the network are usually handled with different methods, such as the configuration of the frequency carriers.

Modern antenna systems are designed to handle frequent modifications in their parameters (e.g. downtilt), without even requiring a visit to the site. It is quite normal to have networks with antennas using the Remote Electrical Tilt (RET) feature, which allows the operator to modify the downtilt angle of the antenna to adapt to new interference conditions. Although far less common, there are also systems to perform Remote Azimuth Steering (RAS), which permits to modify the azimuth pointing angle orientation of the antenna.

Furthermore, many UMTS NodeBs permit to modify the total output power of the amplifier through network parameters, which adds an additional degree of flexibility to the operator who may want to deploy the network in different stages using different sector output power. For instance, the operator may decide to use high power amplifiers (PAs) for rural sites, and low power PAs for urban and suburban environments.

While the adjustment of RF parameters has been simplified in CDMA networks, it is still a complicated, multidimensional task to accomplish. In order to simplify the process, many CDMA operators make use of Automated Cell Planning (ACP) tools. These tools can be considered analogous to the Automated Frequency Planning (AFP) tools used in GSM and other TDMA networks, in the sense that network parameters will require periodic retuning to be optimally configured to the changing traffic conditions. Section 5.4.2 introduces the concept of Automatic Cell Planning.

5.4.1 Practical Rules for UMTS/HSPA Cell Planning

Below is a set of practical rules to be considered when deploying a UMTS/HSPA network:

- pilot power should be adjusted to the specific network conditions;
- limit cell footprints after achieving the coverage goal;
- place sites close to the traffic sources;
- avoid cell splitting for capacity growth in interference-limited networks;
- minimize soft-handover areas;
- consider the effect of HSDPA in overall interference;
- use indoor cells to cover in-building traffic.

These topics will be discussed in detail in the following subsections.

5.4.1.1 Pilot Power Planning

The power assigned to the common control channels, and in particular to the pilot channel, will have a strong influence on the cell coverage. In a CDMA network, it is possible to establish trade-offs between cell coverage and capacity by modifying the power assigned to the common control channels. In dense traffic areas the power assigned to the pilot should be minimized to leave room for traffic channel power, while in areas with small population (for instance, in rural areas) the pilot share can be increased to allow the UMTS pilot signal to propagate further in distance.

5.4.1.2 Interference Control

The interference should be kept well confined within the sector that is creating it. The power from neighboring cells will directly affect adjacent sectors and will reduce their capacity/coverage. Boomer site deployments are generally a bad idea, at least if there are sectors transmitting at the same frequency channel under the boomer's coverage area.

5.4.1.3 Site Selection

The geographical distribution of the traffic is critical in the network design. As mentioned before, the soft capacity of a UMTS cell is highly dependent on where the traffic comes from. The intimate relationship between power and capacity demands that cell sites are placed near to where the bulk of the users are located to achieve higher overall network capacity.

5.4.1.4 Capacity Growth Management

The operator should be cautious when adding new cell sites for capacity reasons in a UMTS cluster, considering the potential of increasing the overall interference level in the existing cell neighborhood. A delicate balance must be found between offloading the traffic from the overload sites and controlling the interference of the new cell. As a general guideline, cell sites should be designed primarily based on coverage and service quality for a certain traffic load, and additional capacity should be provided through additional carriers.

5.4.1.5 Soft-Handover Planning

UMTS networks require a certain overlap between sectors, which will benefit the performance of voice users, however, as explained in Section 5.1.3, too much overlap will hurt the performance of HSDPA users. Operators deploying HSDPA should aim at a soft-handover area below 40% (20% would be a good target for data only networks).

5.4.1.6 HSDPA Interference Considerations

The specified target load for HSDPA cells should be different from that for Rel.'99 only traffic. While sectors with voice will be typically designed to achieve a 70 or 80% load, HSDPA cells can increase power levels close to 100% (depending on operator's configuration). The effect of the increased interference level should be accounted for in the neighboring sites as well.

5.4.1.7 In-building Coverage Planning

In dense urban areas with tall buildings, the coverage inside the buildings should be studied in a different exercise than that of the outdoor planning. It is not a good idea to cover buildings with signals coming from the exterior since the excessive power needed to cover in-building users will create harmful interference to the outdoor users and force those users to raise their power level to maintain the call quality. In the end, it is the indoor users who will lose the battle because they are in less favorable radio conditions, with higher pathlosses to the serving sectors. In addition, the extra power needed to cover the indoor users means less capacity for the cell.

5.4.2 Automate Cell Planning (ACP) Tool Usage

The optimization goal of the ACP tool is to achieve network coverage and capacity targets with the minimized deployment costs. The tools recommend the configurations of the basic radio parameters needed to begin the network deployment, such as the following:

- location of the site;
- height of the antennas;
- antenna types (pattern);

- azimuth orientations of the sector antennas;
- antenna downtilts (mechanical and electrical);
- pilot powers;
- HSDPA max transmit powers.

The ACP tools are typically based on a heuristic engine embedded within a Monte Carlo simulator. The heuristic engine computes a cost function based on a snapshot of the current network configuration, with the associated costs assigned by the operator to each alternative. The cost to add a new site, for example, should be much higher than the cost to modify an electrical tilt of an antenna. The tool has a smart engine to discard the configurations which incur too much cost and then converge towards a solution with minimal cost. ACP tools typically following a mathematical process known as *simulated annealing*. This process can be lengthy, and therefore an important difference among ACP tools is the time they take to converge to an optimal configuration. It is important that the tool has specific algorithms to plan HSDPA traffic. The cost function model should be different from the typical UMTS cost functions to ensure a good HSDPA throughput at the cell edge with minimal impact to Rel.'99 soft-handover performance [15].

An embedded Monte Carlo engine is typically used to estimate the coverage and capacity of the network during the iterations of the optimization process. Monte Carlo simulations provide a detailed performance assessment for the network, both at the geographical bin and sector levels. The simulation results indicate whether the current network configuration is acceptable to carry the desired load, while meeting the operator's criteria for the cell edge coverage and quality of service requirements.

Figure 5.6 illustrates the results of an ACP optimization. The figures display a thermal map of DL throughputs, before and after the optimization, that in the original figure ranged from green (>3 Mbps) to dark red (<600 kbps). Although difficult to distinguish in grayscale, the result of the optimization significantly increases the areas with higher throughputs, and eliminates the coverage holes.

The accuracy of the output of the ACP tool depends on the input data and how the engineer uses the tool. It is very important to gather all the relevant data before proceeding to the planning process. The following are the general 'homework' steps that should be taken by the operator before the ACP process can be initiated:

- Create a table with target performance (QoS) requirements:
  o assign confidence levels for each of the services to be offered;
  o compute a link budget for each service to be offered.
- Gather information about the geographical traffic distribution in the area:
  o traffic maps from legacy network statistics or modeled based on Geographical Information System (GIS) clutter categories.
- Tune the propagation model for the area:
  o use model parameters from drive test tuning or from the research literature for similar area types.

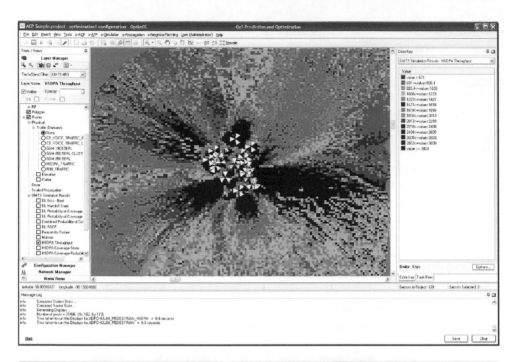

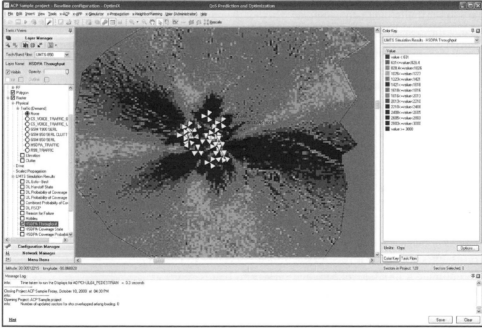

Figure 5.6 Illustration of ACP optimization: HSDPA throughput before (above) and after (below)

- Create a list of possible site candidates for UMTS/HSPA:
  o use existing legacy network sites or siting agreements (e.g. properties owned/leased by the operator or partners).
- Decide the types of antennas that can be used:
  o gather 3D radiation patterns from antenna vendors.
- Define a table with sensible costs ($, €, £ or relevant local currency) per configuration change (e.g. cost per site visit, cost per installation of new antenna, etc.):
  o relative costs are more important than absolute for the purposes of optimization through ACP planning.

Because the traffic is not uniformly distributed in the networks, it is critical to use appropriate geographical traffic maps if possible. In case these are not available, the tool can use information about the clutter type (forest, streets, residential, commercial, etc.) from Geographical Information System (GIS) data to generate traffic. This alternative method is typically less accurate and more difficult to configure because the operator needs to estimate the traffic weights for each clutter type.

If available, drive test measurement data can be used to improve the accuracy of the Monte Carlo simulator, because the prediction data from propagation tools typically has large error margins. In some cases, the amount of drive test data may not be large enough and some locations may not have drive test data available. Some tools have a special feature to combine drive tests with propagation data to tune the propagation model parameters, as shown in Figure 5.7. Note that for initial UMTS deployment the drive test data can be collected from a 2G network and scaled according to the frequency band and antenna pattern configuration for 3G.

After all the input data has been gathered and the success criteria have been properly defined, the ACP process can begin.

5.4.3 Deployment of ACP Network Configuration

After the optimization process is complete, the ACP provides the recommended configuration to be implemented in the network. In the case of optimization of a network that is

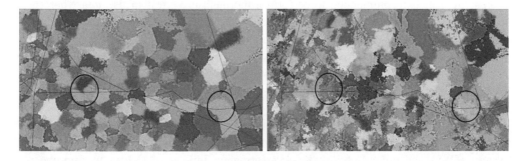

Figure 5.7   Best server plot based on propagation (left) and after combination with drive test data (right)

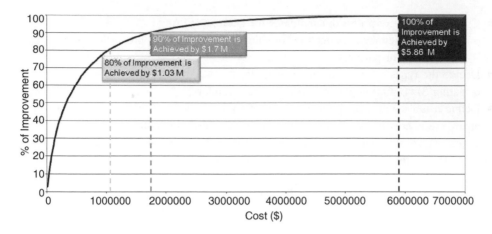

Figure 5.8   Analysis of RF planning cost vs. overall performance improvement

already on the air, it is very important that the sequential steps of changes are provided in order that every incremental change will not degrade the quality of the network. This has an additional benefit in terms of operational costs, since the operator may decide to implement only a few of the changes proposed by the tool. Figure 5.8 illustrates this concept through a Return on Investment (ROI) curve. In this particular example, all the changes required to fully optimize the network (i.e. achieve all the target design goals) would be $5.86M. However, it is possible to achieve 90% of improvement with only a subset of the overall changes with an overall cost of $1.7M. In other words, achieving the remaining 10% improvement would cost more than twice the optimization costs so far, and the operator may decide not to pursue those changes.

UMTS/HSPA ACP tools usually provide software utilities to generate neighbor lists and scrambling code plans, which can be very useful to avoid extensive manual labor and thus decrease the likelihood of human errors.

After the implementation of the ACP network changes, the operator should closely monitor the performance from switch statistics to ensure that the changes have been properly implemented. It is recommended to perform a thorough drive test in the area where new changes have been made and compare the main RF metrics against the results predicted by the ACP tool. Minor fine tuning may be necessary after this step, the details of which are presented in Section 5.5. The actual performance of the services may not be ideal at this point, because they can be related to the configuration of the other network parameters, which are covered in Section 5.6. The software tools for optimizing settings for these other network parameters are discussed in Section 5.7.

When the network is at a mature state where there is sufficient traffic to generate meaningful performance statistics, the ACP process can utilize statistics from the Operational Support System (OSS) and other network Key Performance Indicators (KPIs) to improve the accuracy of the process. Information such as real traffic, pilot distributions, and data application mix, among others, can be very valuable in refining the optimization process over time. In general, real network statistical performance data are more accurate than any predictions or assumptions made in their absence.

5.5 Optimization with Drive Test Tools

As noted in the previous section, after the results of an ACP planning process are implemented the operator should proceed to validate the results of the changes in the network. The most effective way is through drive tests in the area where changes have been made. This subsection describes drive test tools, methods and data interpretation.

There can be one or more drive tests of an area, depending on the level of maturity of the network. For a green field deployment, it is likely that the initial drive tests will be focused on the basic radio parameters to ensure that the coverage of the area matches the predictions from the planning tool. At this early phase of the network optimization, the operator should primarily utilize the following drive test measurements:

- **Common pilot channel (CPICH) Received Signal Code Power (RSCP):** identifies coverage holes where the pilot RSCP is less than a threshold target, for example -110 dBm.
- **Unloaded common pilot channel Ec/No measured in idle mode:** identifies the presence of 'pilot pollution', i.e., too many pilots without a dominant server. Pilot pollution causes quality degradation and capacity loss for Rel.'99 traffic. For HSDPA, the situation become even worse due to constant HSDPA cell change being triggered in those areas. This leads to low throughput and frequent interruptions during the data transaction. For a lightly loaded network, low levels of Ec/No (below -12 dB) should not occur frequently inside the main coverage area. Healthy levels of Ec/No for good RSCP received powers (above -95 dBm) in unloaded conditions should be on the order of -4 to -7 dB.
- **Average number of cells in the active set:** indicates the cell overlap between neighboring cells, which for HSDPA networks should be kept relatively small.
- **Strength of RSCPs in the neighboring cells:** helps to refine the neighbor lists and identify missing neighbors. It is important to capture any cell with a strong pilot in an area that has not been included in the active set.
- **Call setup success rate:** assesses the overall Rel.'99 call quality for mixed voice and data networks. This metric is not purely related to RF because other parts of the network, especially the core network, can be responsible for call failures.
- **CQI distribution for a single user drive:** indicates the radio conditions perceived by the HSDPA device, permitting the operator to estimate the achievable throughputs over the air interface. Earlier discussion noted the handset vendor differences in CQI reporting, so care must be taken in interpreting the results.

Typically, drive data are collected using an external antenna mounted on the top of the drive test vehicle. The measurements collected in this manner represent the outdoor 'on street'

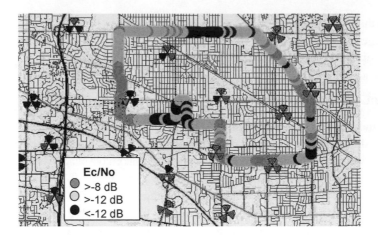

Figure 5.9   Radio conditions (Ec/No) in a cluster from a drive test measurement

values. To estimate the signal strength inside a car, $6 \sim 8$ dB penetration loss needs to be added on top of the collected RSCP. For in-building coverage, $10 \sim 20$ dB building penetration loss has to be subtracted from the outdoor measurements [1].

As an example, Figure 5.9 depicts drive test results for a network prior to optimization. Three problem areas can be identified with RF coverage problems due to low Ec/No shown as dark dots on the drive map.

If no major problems are found from the unloaded drive test measurements, then it is recommended to repeat the exercise under traffic loaded conditions. Orthogonal Channel Noise Source (OCNS) simulators are normally used to generate controllable traffic load; OCNS can be configured on the RNC to load the network to a particular level, for example 60%. This allows the engineers to analyze the performance of the network under loaded condition. One of the drawbacks of OCNS is that only downlink load can be generated. To generate load on the uplink, real traffic could be used. One method is to have multiple Rel.'99 uplink data channels activated during the drive test.

Once the basic RF configuration is verified, the operator has to look into second-level KPIs, beyond the first-level ones listed above, such as data throughput and latency, to better understand the end-to-end network performance. Some recommended follow-up tests are as follows:

- Packet Switched (PS) call establishment time;
- single user, stationary download and upload of large FTP files (larger than 4 MB) under different radio conditions;
- single user with mobility, download and upload of large FTP files during a cluster drive;
- IP pings with 32 byte packets from different sectors in the cluster to analyze the latency;
- joint voice and data tests in a cluster.

The main KPIs to analyze in this second phase are:

- **Total packet switched establishment time:** this can be broken down into System Information Block (SIB) acquisition time, RRC establishment time, PDP context establishment time and Radio Bearer Reconfiguration Time. An optimum setup time (including all the above procedures) should be around 2 seconds.
- **Application level throughput:** in good radio conditions, the bitrate should be limited by the device capability, NodeB baseband configuration and Iub backhaul resources. A bitrate lower than expected by a margin of 10% would likely indicate an issue worthy of troubleshooting.
- **Latency (Round Trip Time):** under ideal conditions, the RTT should be close to 90 ms for HSDPA/Rel.'99 DCH, and 70 ms for HSDPA/HSUPA. In practice, this RTT will be increased by the transport delay (a few milliseconds, depending on the location of the SGSN and GGSN). Round trip times longer than 150 ms indicate an issue to troubleshoot.
- **Packet drop call rate in stationary conditions:** packet drops in good radio conditions indicate a problem.
- **Packet drop call rate during cell reselections:** packet drops in good radio conditions indicate a problem.
- **Throughput reduction or interruption:** particularly duration of the effects during cell reselections.
- **Voice performance:** assess voice quality when data services are present (BLER, drops, access failures).

It is common that not all of these follow-up tests yield good results after the network has been initially deployed. The fact that radio, core and transport effects are coupled together can make it difficult to troubleshoot a particular problem. The operator may have to dig into each of the problems and try to resolve them from an end-to-end perspective. The following are a few useful troubleshooting tips for resolving typical issues found in HSPA networks based on our experiences:

- **High latency:** do not underestimate latency problems. Increased network latency will affect the application level throughput and impact the end-user experience. Typical reasons for increased latency can be:
  - configuration of state transition timers;
  - GPRS core architecture with long delays between RNC and GGSN;
  - configuration of SGSN;
  - location and stability of the application server; and
  - congestion in any of the network paths or nodes.
- **Packet drops:** in good radio conditions these suggest problems in the core or transport network. The Drop Call Rate (DCR) in these cases should be zero or close to zero.

- **Reduced throughput:** in good radio conditions, if the latency measurements are good then this will likely be a configuration problem (radio or transport) or an issue with the FTP server. Note that sometimes the TCP parameters in the server and/or the device have to be modified to achieve the highest throughputs (see Chapter 3 for further details on TCP/IP parameter settings).
- **Intra-RNC mobility problems:** these suggest either dimensioning issues or radio configuration issues in the RAN.

As an example of a follow-up drive to collect second-level KPIs, Figure 5.10 illustrates an example of a HSDPA cluster drive. The mobile device utilized during the drive test was a PC data card supporting HSDPA Category 6. The network was configured to use 5 spreading codes, 16 QAM modulation and 2xT1s backhaul per NodeB. The maximum throughput should have been in the order of 2–2.3 Mpbs (limited by the backhaul); however, the results indicate a maximum bitrate of 1.4 Mbps. In this case the problem was determined to be the Iub flow control configuration, which was inadvertently set to a maximum of 1.6 Mbps. Also, the plot on the right illustrates the performance of cell reselections during the drive, with visible throughput degradations during the transitions (the deep dips or valleys in the throughput). Such performance may be normal depending on infrastructure vendors' implementation, since the MAC-hs buffers are being flushed every time when there is a change in the serving NodeB (see Chapter 4 for more details on this phenomenon). In other cases it may indicate parameter configuration issues related to the direct HS-DSCH cell change.

Many problems found during drive testing in the network are related to the configuration of radio and transport parameters for HSDPA and HSUPA. Although the infrastructure vendors' default parameter settings are typically adequate for the basic operation of the network, in practice it is necessary to optimize many of them to make sure they are optimum for the specific network deployment. Section 5.6 discusses the main parameters related to the operation of HSPA and their corresponding tradeoffs from an operator's point of view.

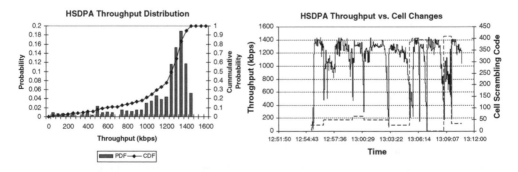

Figure 5.10 Example of follow-up HSDPA drive test to obtain second-level KPIs to measure performance

5.6 Main Radio Parameters Affecting HSPA Performance

This section reviews the main radio parameters affecting the performance and capacity of HSDPA and HSUPA. Although all infrastructure vendors provide a set of default parameters as a starting point for the operation of the network, the operator will have to adjust some of these to their specific situation, such as target traffic type, expected data load, last-mile transport provisioning and most importantly, available budget for deployment.

Every infrastructure vendor has a different set of parameters which typically refer to the same particular feature defined in the standard. This section will not cover specific vendor implementations, but instead refer to the abstract concepts that are applicable across different manufacturers.

The parameter configuration can be implemented through a user interface in the corresponding element manager, through a particular license that needs to be loaded, or through constant fields that need to be specified by external parameter files. See Chapter 4 for more details on the algorithms and functionalities controlled by the different parameters.

Table 5.11 provides a condensed summary of the HSPA parameters described in later subsections, including the advantages and disadvantages that can be attained by modifying such parameters.

5.6.1 Basic Activation Features

5.6.1.1 Enable HSDPA

Typically the operator has the choice of activating HSDPA on a per cell basis. Given the huge performance and capacity improvements by HSDPA, the normal choice would be to enable the functionality; however there are occasions in which it may be better to disable it, such as the following:

- **In cells with very little data traffic that are limited in baseband resources.** Since the activation of HSDPA requires a certain upfront reservation of channel elements in the baseband resources, this may require additional hardware in the BTS. If the traffic data is minimal in the cell, these costs can be saved by leaving the data on Rel.'99 channel. It should be noted that this would typically be an interim solution because data demand is likely to grow and eventually the operator would be better off serving the larger data traffic with HSDPA.
- **At the border of UMTS coverage areas.** As discussed before, due to the lack of compressed mode support for HSDPA, it will not be possible to perform direct HSDPA to (E)GPRS cell change when the UMTS coverage ends. While in principle it should be possible to transition the calls from HSDPA to Rel.'99 within the same cell, in practice this method has been proven to be ineffective or difficult to adjust and causes call drops on HSDPA. By having a buffer area where no HSDPA traffic is allowed, the mobiles will smoothly transition to the DCH channel during the cell transition into the buffer area, from which it is possible to perform a normal 3G to 2G handover.

Table 5.11   Overview of principal HSPA parameters

Category	Parameter name	Advantage	Disadvantages
Basic activation features	Enable HSDPA	Improved DL throughput, latency & capacity	Baseband cost IRAT mobility issues
	Enable HSUPA	Improved UL throughput, latency & capacity	Baseband cost
	Enable 16 QAM	Improved DL throughput and capacity	n/a
Control of resources	Enable HSDPA Dynamic code allocation	Improved DL throughput, minimized RT blocking	Baseband cost
	HSDPA Code selection	–	–
	Number of HS-SCCHs	Improved capacity for low-bitrate applications	Increased code and power consumption
	Enable HSDPA Dynamic flow control	Improved efficiency in mixed voice/data carriers	n/a
	Static Iub flow bitrate	–	Requires specific configuration based on Iub resources
	Enable HSDPA Dynamic power allocation	Improved DL throughput and HSDPA coverage	Possible effect on voice quality
	HSDPA power	–	–
	Number of HSDPA users	Increased HSDPA capacity	Baseband cost
	Maximum noise rise with HSUPA	Increased HSUPA capacity and user bitrates	Increased UL interference
	Number of HSUPA users	Increased HSUPA capacity	Baseband cost
	HSUPA code selection	Higher UL bitrates	Increased UL interference
Mobility management	Enable cell selection to HSDPA layer	Useful for HSDPA redirection with multiple carriers	
	Enable Direct HSDPA Cell change	Improved average throughputs in cell transitions, in creased capacity	Possible impact on RT packet switched services
	HSDPA Serving Cell Change triggers	–	–
	HSDPA Serving Cell Change Hysteresis	Help solve ping-pong effects	Can degrade perfor- mance if set too large

Table 5.11   *(Continued)*

Category	Parameter name	Advantage	Disadvantages
	HSDPA transition to DCH	Useful for transitions to non-HSDPA coverage area	Degraded throughput and capacity
	Max HSUPA bitrate during soft handover	Limits interference and consumption of baseband resources	Degraded UL throughput
Performance	HSDPA Scheduler type	Improved performance with PFS	PFS does not provide maximum cell capacity
	HARQ Type for HSDPA	Improved performance with IR	n/a
	HARQ Type for HSUPA	Improved performance with IR	n/a
	Enable CQI adaptation	Improves link adaptation performance	n/a
	Max HSDPA or HSUPA bitrate	Help control resource utilization	n/a
	UL bitrate with HSDPA	High values can improve latency	Possible capacity issues
	FACH to PCH Transition	Large timers will improve latency	Possible capacity issues
	FACH to DCH Transition	Short timers and thresholds can improve user experience	Possible ping pong effects with bursty data
	DCH to FACH Transition	Long timers will improve user experience	Possible capacity issues
	Channel Transition between HS-DSCH and Rel'99 DCH	Short timers will improve user experience	Possible ping pong effects
	HSUPA Fast Power control parameters	Large steps will improve UL bitrates in low loaded conditions	Possible impact in sector capacity
	HSUPA Scheduler parameters	–	–
	Number of HARQ retransmissions	Large values improve coverage and spectral efficiency	Latency suffers with large number of retransmissions
	Selection of HSUPA TTI	Improved latency and peak throughput with 2ms TTI	Possible coverage issues with 2ms TTI

5.6.1.2 Enable HSUPA

HSUPA can be activated through a cell level parameter. Because the majority of the data traffic is on the downlink due to asymmetrical internet data traffic model and the baseband resource required to activate HSUPA could be substantial, the operator may decide to delay the activation of HSUPA until the additional bitrate or capacity is necessary in the cell. The addition of HSUPA will have a significant improvement to end-to-end latency. With applications such as video sharing (e.g. see what I see, mobile gaming) becoming more and more popular, the demand on uplink data capacity will eventually catch up, the operators can selectively activate HSUPA in areas where high uplink capacity is in demand.

5.6.1.3 Enable 16 QAM Modulation

This parameter controls the activation of the 16 QAM modulation, which enables the user to achieve higher bitrates in good radio conditions. While it has no major drawbacks, the activation of 16 QAM has an important effect on user peak throughput as well as in overall sector capacity. In some implementations, 16 QAM is active by default and cannot be deactivated. In any case, the operator should check that the feature is activated, either through parameter or license file; otherwise the cell will not achieve the expected maximum throughput performance.

5.6.2 Control of Resources

5.6.2.1 Enable HSDPA Dynamic Code Allocation

The code allocation in a cell can be static (a fixed number of codes for HSDPA is selected) or dynamic. When this parameter is turned on the system will be more efficient because the capacity for HSDPA will adapt to the capacity left by the real-time services. On the other hand, the simple activation of this parameter may require a larger upfront reservation of baseband resources and therefore may not be recommendable in sectors with light load. To limit this impact, in some implementations it is possible to configure the maximum number of codes to be used by this feature.

5.6.2.2 HSDPA Code Selection

For both dynamic and static code allocation, it is possible to select the number of codes to be used by HSDPA. This selection will be performed either based on a maximum number of codes, or on a specific code set determined by the operator. There are code allocation algorithms that also provide a configurable margin for Rel.'99 traffic in order to avoid congestion of real-time services. It is recommended to adjust these parameters to the expected air interface load and planned Iub backhaul capacity to ensure cost effective utilization of the resources. For instance, if only 1xT1 is configured there is no point in configuring more than five codes in the Node-B.

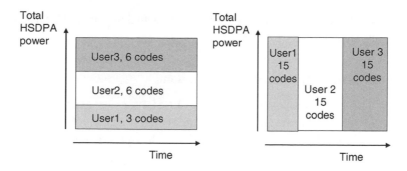

Figure 5.11   Illustration of TTI multiplexing (left, 3 HS-SCCH) vs. no multiplexing (right, 1 HS-SCCH)

5.6.2.3 Number of High Speed – Shared Control Channels (HS-SCCHs)

The number of High Speed – Shared Control Channel (HS-SCCHs) determines the code multiplexing capabilities of the sector within a TTI. The number indicated by this parameter determines how many users may transmit at the same time in the same TTI. Figure 5.11 illustrates this concept, with the case on the left showing a cell configured with three HS-SCCHs (three users can be code multiplexed), and the right figure showing a configuration with only one HS-SCCH.

Setting this parameter to a value higher than one is useful for sectors that serve many simultaneous HSDPA users with low-bitrate applications such as Voice over IP (VoIP), however, in most of the cases there is no need to use more than one HS-SCCH. Furthermore, each of these codes consumes additional power resources from the sector; therefore they should not be activated if they are not going to be used. The activation can also have an impact on the baseband resources of the cell, due to the fact that some implementations only permit the use of this feature when all 15 codes are allocated to HSDPA.

5.6.2.4 Enable HSDPA Dynamic Flow Control

The flow control mechanism for HSDPA manages the allocation of transport resources between the RNC and the NodeB. It is an important mechanism because the RNC is not aware of the transmissions performed at the physical or MAC-d layer. It is the NodeB which must indicate to the RNC how many more frames it can handle at every instant in time.

This parameter selects the dynamic flow control, which performs an automatic adaptation of HSDPA bitrate based on the available Iub backhaul bandwidth. The feature is typically controlled by associated parameters that determine when to increase or decrease the bitrate based on current demand and overall congestion situation.

5.6.2.5 Iub HSDPA Bitrate for Static Flow Control

In the case of static flow control the operator must define the maximum Iub bitrate to be allocated for HSDPA traffic. This parameter must be carefully configured based on the provisioned Iub resources because it will determine the actual maximum HSDPA bitrate to be delivered by the sectors in the NodeB.

5.6.2.6 Enable HSDPA Dynamic Power Allocation

As covered in the previous chapter, the allocation of HSDPA power can be either static (fixed amount of power allocated to HSDPA) or dynamic (HSDPA and Rel.'99 traffic share all the available sector power). It is recommended to activate the dynamic power allocation feature because it provides a more efficient utilization of the resources, which leads to a better data experience, especially at the edge of the cell.

5.6.2.7 HSDPA Power (Static Allocation)

This parameter defines the amount of power to be used by HSDPA in the case of static power allocation. Typically this power will be allocated when the first HSDPA users start transmitting, and will be fully de-allocated in the case of Rel.'99 congestion in the cell, or in the absence of HSDPA traffic. Moderate values are recommended (e.g., 30% of total power, such as 6 Watts for a 20 Watt Power Amplifier), otherwise the HSDPA service may have to be released too frequently due to congestion situations, which will impact user performance.

5.6.2.8 HSDPA Power (Dynamic Allocation)

In the case of dynamic power allocation, some implementations permit the operator to define a maximum sector power to be used when HSDPA service is allocated, while other services will utilize all the remaining power. Using all the remaining power for HSDPA may result in shrinking coverage and potentially degraded voice service under certain network conditions. Therefore setting up a limit on the HSDPA power makes sense, especially when the data traffic is not high and the operator wants to maintain good voice quality.

The maximum power for HSDPA can be defined in absolute terms or relative to the current cell power utilization. Our recommendation is to set the maximum HSDPA power to about 40% of the maximum, for example 8 Watts for a 20 Watt power amplifier in the case of shared voice and data carriers.

5.6.2.9 Admission Control Parameters with HSDPA

In some implementations it is possible to define different sets of admission control parameters when HSDPA traffic is present in the sector. This is useful to prevent blocking of traffic due to high HSDPA transmit power. This strategy can be adapted to admit other types of traffic with higher priority than HSDPA. It is recommended to set these parameters to a more aggressive load value when HSDPA traffic is present to achieve a higher efficiency in the utilization of network resources.

5.6.2.10 Maximum Number of HSDPA Users

This parameter indicates the maximum number of simultaneous users allowed in the cell. While having a large number of users is desirable from a capacity perspective, this would require a larger upfront reservation of baseband resources, therefore the operator may want to set a lower number of users in cells with low data traffic volume.

5.6.2.11 Maximum Noise Rise with HSUPA

This parameter determines the maximum overall uplink noise rise permitted in the cell when HSUPA is present. Vendors implement this differently to be either noise rise with HSUPA power included or not. A recommended value is 10 dB if it includes the HSUPA power or 4 dB if HSUPA power is not taken into account.

5.6.2.12 Maximum HSUPA Uplink Noise

In some implementations it is possible to control the amount of HSUPA power contributing to the overall noise rise. Conservative values will result in lower interference in the cell, but also reduced uplink capacity and lower user bitrates. In cells with mixed voice and data it is not recommended to assign a large value to this parameter. A compromise setting to achieve high single user's speed and reasonable sector capacity is around 6 dB. In sectors with data only, this value can be increased to 10 dB or even as high as 30 dB to achieve the best user experience.

5.6.2.13 Maximum Number of HSUPA Users

Similar to the *maximum number of HSDPA users*, this parameter controls the number of users that are served by the scheduler. In cells with low data utilization it is recommended to set this parameter to the value that requires less baseband resources, typically less than four.

5.6.2.14 Selection of HSUPA Codes

The operator typically has the option to define the spreading factor to use (SF2, SF4 or both) and/or the number of parallel codes (one or two). These parameters help control the resources consumed by HSUPA and establish a tradeoff between user performance and overall network interference.

Using SF4 provides better performance than SF2 by enabling higher processing gain. SF2 can be used in very good radio conditions with a much higher bitrate. However, higher bitrate will demand more baseband resources and a higher uplink transmit power which can lead to rapid uplink noise rise in the cell. Similarly, using two codes effectively doubles the user bitrate, but also increases the consumption of baseband resources and transmit-power by a factor of two.

Because the interference level in the cell can be controlled through various radio resource management algorithms such as congestion control and UL packet scheduler, it is not recommended to use these configuration parameters for interference control purposes.

5.6.3 Mobility Management Parameters

The mobility management parameters in this section are primarily related to HSDPA. Although HSUPA uses soft-handover procedures for the uplink, the serving cell changes typically follows that of HSDPA.

5.6.3.1 Enable Cell Selection to HSDPA Layer

Enabling this parameter will transfer HSDPA capable mobiles to a preferred HSDPA carrier frequency. This is useful for situations in which the operator has deployed several carriers and wants to keep data traffic separated from voice as much as possible. This scenario has some advantages with regards to service quality and impact to voice users, as are discussed in Chapter 7.

5.6.3.2 Enable Direct HSDPA Cell Change

This parameter activates the High Speed-Downlink Shared Channel (HS-DSCH) cell change procedure, which permits direct cell reselections of data calls in the HS-DSCH channel. If this feature is not active, HSDPA users will fall back to Rel.'99 channel during the cell transition and will eventually come back to the HSDPA channel in the target cell. It is generally recommended to use direct HS-DSCH cell change because transitions to Rel.'99 will likely result in a slower data rate and the user perception may be degraded. Also, if the number of HDSPA users in the sector is high, performing frequent transitions to Rel.'99 may result in congestion situations and possible drops.

On the other hand, there are situations where it may be advantageous to disable HS-DSCH cell change, such as in RNC boundary cells where HS-DSCH cell change can have relatively poor performance.

5.6.3.3 HSDPA Serving Cell Change Triggers

These parameters define one or more criteria that need to be satisfied before a HS-DSCH cell change can occur. The cell change trigger can be defined in terms of RSCP, Ec/No or uplink performance. It is preferable to use the RSCP criteria because Ec/No will vary with load, and may change abruptly when dynamic power allocation is activated for HSDPA.

The event that triggers the cell change is called event 1d-hs, which happens when a cell in the active set of Associated Dedicated Channel (A-DCH) meets the criteria of replacing the current HSDPA serving cell, or the serving cell leaves the A-DCH active set. Therefore it is important to review the soft-handover parameters for the A-DCH.

5.6.3.4 HSDPA Serving Cell Change Hysteresis

This parameter is extremely important because it can help limit the amount of HS-DSCH cell changes during a single connection. As discussed in previous chapters, HSDPA performance can suffer significantly during cell change and they should be carefully controlled. The hysteresis parameters typically provide a delta value that the new serving cell has to satisfy. This ensures that the mobile will not ping-pong between the current serving cell and the target cell. Continuous ping-pong effects may trigger a transition of the HSDPA user to a Rel.'99 channel which is also undesirable. However, an excessively large hysteresis value can result in performance degradation because the delayed cell change may cause the interference received by the mobile to be too strong and lead to call drop.

In some implementations, the hysteresis condition needs to be satisfied for a certain time before the procedure is actually triggered. This triggering time is an associated parameter that needs to be configured and similar tradeoff needs to be made with it.

5.6.3.5 HSDPA Transition to DCH

There are two main reasons based on which the system may downgrade an existing HSDPA connection to a Rel.'99 channel: poor signal quality and handover performance at HSDPA cluster boundary. While in general it is not recommended to force a downgrade to DCH, this mechanism is useful to handle HSDPA transitions to areas where there is no HSDPA coverage, either in UMTS boundary areas or in HSDPA hotspots. For the first case, an absolute RSCP value may be used to trigger the handover, while in the second case it's recommended to use relative RSCP comparisons. For both cases to work, the cell must have disabled the HSDPA direct cell change feature, therefore when the handover criteria is met the mobile will perform a channel switching from HS-DSCH to DCH.

5.6.3.6 Max HSUPA Bitrate During Soft-handover

This parameter determines the maximum bitrate to be used by HSUPA calls when in soft-handover state. Limiting this value will help reduce excessive consumption of cell resources in multiple cells at the same time, and will limit the interference in the non-serving cells which don't have direct control on the uplink power. On the other hand, a very low value will impact the quality of the data service. In some implementations it is possible to specify a bitrate limit beyond which the *non-serving cells* will not process the calls. This is an effective way to limit the resource consumption in the non-serving cells while not impacting the user data rate.

5.6.4 Performance Parameters

5.6.4.1 HSDPA Scheduler Type

As we reviewed in Chapter 4, there are several types of packet schedulers that can be used with HSDPA; the most commonly implemented ones are Round Robin, Max C/I and Proportional Fair. The selection of the scheduling strategy has a significant impact on the cell capacity and user experience. Round Robin schedulers will not achieve the maximum capacity, but will provide the same amount of turns to all users in the cell; it is typically not recommended but can be useful for certain cell types such as indoor locations. The Max C/I schedulers maximize the spectral efficiency of the sectors at the expense of users in poor conditions; therefore it is not recommended if the operator wants to offer a similar user experience to all users, more or less independently of their location. The Proportional Fair scheduler is the most accepted option because it provides capacity gains compared to the Round Robin, while ensuring a relatively fair treatment to all users in the sector. Our recommendation is to use Proportional Fair scheduler as a general rule; however the operator may want to consider Round Robin for special cases where the CQI report may not be very accurate, such as indoor scenarios.

5.6.4.2 HARQ Type for HSDPA

The MAC-hs retransmission mechanism can be Chase Combining or Incremental Redundancy (IR). Both methods perform combination of retransmitted frames. IR is more versatile because it allows the retransmissions to be different from the original packet. It is therefore recommended to use IR as the HARQ method.

5.6.4.3 HARQ Type for HSUPA

Similar to the HSDPA case, it is also possible to select the combining method for HSUPA HARQ. As in the previous case, it is recommended to use Incremental Redundancy over Chase Combining.

5.6.4.4 Enable CQI Adaptation

Some link adaptation algorithms have an internal mechanism to adapt the CQI reported by the mobile based on the actual DL error rate. This feature is very useful because the CQI values reported by different mobiles can vary significantly from each other, and therefore the link adaptation algorithm will not work properly if it is guided by an inaccurate CQI value. It is recommended to turn this feature on to let the NodeB modify the reported CQI to a value where the link adaptation operates most efficiently.

5.6.4.5 Max HSDPA or HSUPA Bitrate

Although the maximum user data rate should normally be defined by the core network Radio Access Bearer (RAB) parameters (negotiated during the PDP context establishment), it is sometimes possible to further cap the bitrate at the radio level. Because there are other methods to control the utilization of resources in the sector, it is recommended to set these parameters to the maximum possible values, otherwise they can lead to degraded performance which will be difficult to troubleshoot.

5.6.4.6 Uplink Bitrate with HSDPA

This parameter defines the initial bitrate that will be used in the uplink when a HSDPA connection is created and HSUPA is not currently active in the sector. The value of the uplink bitrate can have an effect on latency, therefore it is not recommended to have a very low uplink bitrate when the connection starts.

5.6.4.7 FACH to PCH Transition

The transition between the Forward Access Control Channel (CELL_FACH) and Paging Channel (CELL_PCH) states is triggered based on the status of the transmission buffer. If the buffer is empty for a certain time, the mobile will be moved to either CELL_PCH or UMTS Registration Area Paging Channel (URA_PCH) state, depending on the operator's choice. In PCH state, the UE does not have a physical link to transmit data, therefore is not consuming sector resources, but any new data transmission will require an additional time to setup the link thus increasing the latency of the first packet. The selection of this parameter has to be performed based on the expected data traffic type and the actual load. A large timer value will improve latency but may create capacity issues in the cell.

5.6.4.8 FACH to DCH Transition

To explain this parameter, Figure 5.12 presents a simplified state transition model. The transition between FACH and DCH states is typically based on a load threshold for the data buffer and an associated timer. If the amount of data in the buffer does not exceed the threshold for the specified time, then the data will be transmitted at a very low speed (typically 8 kbps). This can seriously degrade the user experience, especially with bursty traffic types. On the other hand, if the data buffer becomes full, then the mobile will undergo a Radio Bearer Reconfiguration procedure which typically is a slow procedure (in the order of 2 seconds), which will again undermine the user experience.

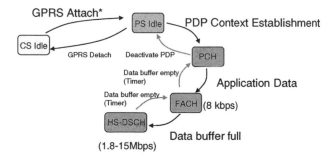

Figure 5.12 State transition model for HSDPA data

5.6.4.9 DCH to FACH Transition

This parameter set should be designed in accordance with the FACH to DCH parameters. In a similar fashion, the transition to FACH would typically be based on the buffer utilization for a certain period of time. The call is moved to FACH through a Radio Bearer Reconfiguration procedure. It is recommended to set these thresholds and timers so that the mobiles that are in transmit mode stay in CELL_DCH state and do not ping-pong back and forth. This parameter will have to be planned based on the typical service type in the network and the available resources.

5.6.4.10 Channel Transition Between HS-DSCH and Rel.'99 DCH

Although in cells with HSDPA the packet data session will normally be on the HS-DSCH channel, there are situations in which the call will be carried on a Rel.'99 channel. In such cases, the cell will try to move the user to the HS-DSCH channel. Sometimes the access to HSDPA is limited by a timer (guard timer) together with a quality threshold that will prevent continuous transitions between Rel.'99 and HSDPA. To improve the customer experience it is recommended that this timer is set to a small value, and the ping pong effect is controlled with the transition criteria to DCH (handover parameter).

5.6.4.11 HSUPA Fast Power Control Parameters

In HSUPA, the uplink power is controlled by the fast power control algorithm and the HSUPA scheduler. For the fast power control, it is possible to define the initial target SIR for the connection. During the communication the SIR will be adapted based on a specified error rate, which may be defined in terms of BLER or the number of HARQ retransmissions. The SIR target will be adjusted up or down based on parameters controlling the SIR step change. Large steps will result in better data performance in low loaded situations, but may impact the overall sector capacity due to the sudden interference jump on the uplink.

5.6.4.12 HSUPA Scheduler Parameters

The main parameters controlling the behavior of the HSUPA scheduler are the different target noise rise parameters, previously introduced in the section covering the control of the resources. Depending on the actual implementation there can be other parameters to control the assignment of resources on individual links, such as the initial serving grant value or a target bitrate for the scheduled users.

5.6.4.13 Number of HARQ Retransmissions

It is possible to define the number of uplink retransmissions. Increasing the number of retransmissions will improve the cell coverage and spectral efficiency of the sector, but it will have a negative effect on latency. Typical values for this parameter are 2 or 3. The value should be selected based on the operator's strategy (capacity vs. user experience) prior to the cell planning process, as it is directly related to the Eb/No required by the connection and will therefore impact the link budget.

5.6.4.14 Selection of HSUPA TTI

With HSUPA, it is possible to select between two different frame lengths, 2 ms and 10 ms. The 2 ms TTI provides better latency and higher peak throughput, while the 10 ms frame is better suited for bad coverage conditions [16]. The selection of this parameter will be based on the capability of UEs distributed by the operator, as well as the RF conditions of the sector. In general it is recommended to use 2 ms TTI except for sectors with coverage problems.

5.7 Dynamic Network Optimization (DNO) Tools

Given the increased complexity of the wireless systems being deployed, with a growing number of features and equally growing number of parameters, it is becoming a difficult task for operators to have an engineering team capable of undertaking the optimization exercise for the overall radio network. In most of the cases, the engineers focus on the main radio metrics and largely apply the default Radio Resource Management (RRM) parameters to all the sectors, leaving substantial room for improvement compared to optimization done on a sector-by-sector basis.

In order to achieve optimum efficiency and performance, the network should ideally be configured on a cell by cell basis and based on the time of the day too, because traffic distributions change dramatically between daytime office and nighttime residential hours [17]. However, in practice it is extremely difficult to design and manage so many changes on a manual basis and would require an incredible amount of engineering resources – and risk human errors due to the complexity. This is one of the reasons why 4G technologies such as LTE include features called Self Organizing Networks (SON).

The concept of Self Organizing or Self Optimizing networks is not 4G specific, because it has been discussed and tried, even in 2G systems [18–20]. Furthermore, some 3G manufacturers offer a set of automatic optimization algorithms embedded in their Operations Support System (OSS), such as automatic neighbor list optimization, optimization of CPICH power, soft-handover thresholds, etc. Additionally there are third-party companies that offer automatic optimization solutions that can connect to the operator's OSS to extract the relevant KPIs and output an optimized set of parameters for that network [21].

Today these automatic optimization techniques have not been widely adopted, primarily due to operators' concerns about losing control of the network by letting an automated tool 'blindly' modify sensitive parameters. A similar scenario occurred when new tools for Automatic Frequency Planning (AFP) were introduced in 2G networks, or Automatic Cell Planning (ACP) started in 3G networks. The new tools had to be proven in many networks before operators became comfortable 'taking their hands off the wheel'. Now both AFP and ACP techniques are widely accepted as standard network operating practices. In the same manner, it is expected that Dynamic Network Optimization (DNO) tools will break through the same barrier and become extensively used in the near future.

Note that the DNO system manages soft network parameters (configurable through the OSS), while ACP deals primarily with RF parameters that require a modification of the hardware (such as antenna azimuth, downtilt, height, etc.). In some cases the DNO could also be used to optimize certain RF parameters, such as pilot power and electrical tilt, if the network is equipped with RET antennas controlled through the OSS.

With a DNO system the operator can gain access to multiple optimization strategies and apply them seamlessly with minimal human interaction – and hence minimal opportunities for human errors. The concept, illustrated in Figure 5.13, relies on the intensive gathering of information by the OSS, which is stored into internal counters and combined into specific Key Performance Indicators (KPIs). The DNO system extracts the relevant statistics and applies a set of predefined rules. These knowledge-based rules mimic the thinking process of an optimization engineer to determine a specific modification of the existing parameter configuration space. As with AFP and ACP, this process is iterative, and will be repeated periodically to slowly converge towards an optimal configuration, adapting along the way to changing traffic demands.

The main advantages of utilizing DNO solutions are:

- DNO techniques provide fast and mathematically optimum configurations of the network.
- Continuous network optimization, adapting to growing traffic demands.
- Permits operators to optimize all aspects of the network at once, reducing optimization time and engineering effort.

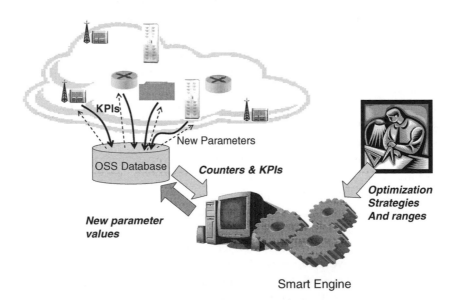

Figure 5.13   Concept of DNO operation

- Optimized each sector independently, ensuring an optimum configuration in each and every location and improving overall network efficiency.
- Permits assigning different configuration strategies for different times of the day (e.g. business busy hour vs. residential busy hour).
- Customized optimization strategies can be applied network-wide or in local markets.

Similar to the optimization processes performed manually today, every iteration of the DNO process is performed in four phases:

1. Gathering of relevant statistics and KPIs.
2. Analysis of KPIs to determine problems and possibilities for improvement.
3. Selection of optimization strategy to be applied depending on the situation.
4. Modification of the relevant parameters according to the optimization strategy.

5.7.1 Collection of Relevant Network Information

One important consideration regarding DNO systems is the fact that the optimization decisions will be based on statistics that are collected automatically by the network. This has the advantage that no additional effort is required by the operator, such as gathering drive test records from a certain cluster, but on the other hand the operator must ensure that the data collected is statistically relevant. This requires that the network be optimized to carry a minimum amount of traffic, making this technique valid only for networks that have reached a certain maturity. For instance, it would make little sense to optimize a sector that has on average two active data calls in the busy hour, simply because these calls could be originated from different locations every day and the optimization result could vary randomly and widely from day to day.

A good optimization exercise needs to gather (1) the most relevant statistics related to the optimization process and (2) as many statistics as needed to avoid statistical noise in the results (i.e., statistically significant results). The first point is solved by defining Key Performance Indicators (KPIs) for each optimization strategy. The second point needs to ensure that the target area carries enough traffic and that the observation window prior to the optimization is long enough to provide sufficient quantities of data. For example, in high traffic sectors the optimization decision could be based on data from one hour or one day, while in lightly loaded sectors the statistics may have to be accumulated for a longer time.

As mentioned above, the KPIs to be collected depend on the particular optimization goal. For example, when optimizing the HSDPA throughput in a sector the most important KPIs would be the overall quality in the sector (CQI and Ec/No distributions), amount of real-time traffic load, the average and maximum number of HSDPA users in the cell, and the typical duration of the PDP contexts, among others. On the other hand, most of these KPIs would be quite irrelevant for an optimization exercise targeting to improve the cell reselection performance with HSDPA.

The optimization algorithms will typically try to find a balance between coverage, capacity and quality. The typical KPIs will provide sufficient information in each of these categories. Some example KPIs are as follows:

- **Quality:** Ec/No, CQI, BLER, physical channel throughput, RLC throughput.
- **Coverage:** pilot RSCP, pilot Ec/No.
- **Capacity:** total cell transmit power, voice and data transmit powers, blocking of real-time traffic, blocking of packet swtiched calls, downgrades of HSPA to Rel.'99.

Because there will be many possible combinations of KPI statistics needed, it is frequent to find that the infrastructure vendor may not have defined the required counters or KPIs. This is normally not a showstopper, because the vendors have processes in place to allow the operator to define new counters; however, this will take time and the volume of customization should be limited. For this reason it is important that the operators clearly define the optimization strategy beforehand and identity the required statistics well in advance, so that they can be ready when the network reaches the right level of maturity.

5.7.2 Identification of Parameters for DNO

In the same way as with the selection of relevant KPIs, before the DNO process a series of relevant parameters for adjustment must be selected. These parameters should be chosen based on their impact to the overall optimization goal: for example, in an optimization exercise designed to tune the maximum HSDPA power, the most important parameters to modify would be the HSDPA power and the admission control thresholds for voice traffic.

Section 5.6 can be used to identify the most relevant parameters in each particular case. Some of the typical parameters to tune with DNO are as follows:

- CPICH transmit power;
- HSDPA maximum transmit power;
- cell reselection hysteresis;
- soft-handover windows;
- hard-handover triggers and thresholds;
- admission control thresholds.

5.7.3 Definition of the DNO Strategy

Once the desired optimization goal has been defined, and the relevant parameters and KPIs have been identified, the operator should define the optimization strategy. The strategy has to take into account different conditions that can be faced by each of the sectors in the network. The strategy should plan how to react to a wide range of possible situations. Instead of creating knowledge-based rules for every single combination of scenarios, the DNO tools have a certain logic embedded that can be as simple as a binary rule or as complex as neural network logic.

Depending on the tool, the operators may have the ability to fully design the optimization strategy, which can capture the expertise of their key optimization engineers into processes that would be applied equally to all the sectors in the network. In other cases, the tool provides the operator with a choice of optimization goals on which the operator can only control a certain part of the process, for instance by assigning weights or costs to the different features of the algorithm.

In general, some practical considerations that are desirable in DNO tools are:

- The tool should provide an embedded parameter consistency check to identify when certain parameters are out of the desired range.
- Different optimization strategies running simultaneously in the same cluster should be coordinated, and should not modify the same parameter without jointly considering the goal of other optimization algorithms.
- Algorithms and thresholds should be accessible and easy to modify by engineers. The optimization strategy should allow a degree of customization because many operators have different goals for their networks.
- The tool should provide ways to let the operator confirm the parameter changes before they are implemented. This is particularly useful during the first iterations to confirm that the optimization strategy is working correctly.

Figure 5.14 presents an example of the execution of a DNO tool in a live 3G network. In this particular case, the optimization strategy was designed to improve dropped call rates. As it can be observed in the chart, the Drop call rate (RAB DR) was significantly reduced during the

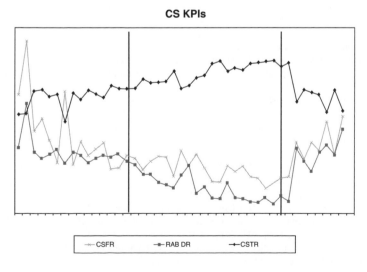

Figure 5.14 Example of execution of an automated parameter optimization (reduction of call failures)

optimization period, and the traffic (CSTR) carried by in the network increased too. After reverting to the previous parameters the performance of the network came back to the previous levels.

5.8 Summary

Below are the main takeaways from this chapter:

- HSPA RF planning requires a special treatment as compared to regular UMTS traffic. We presented some practical rules to account for HSPA specific needs in the planning process.
- The link budget analyses for HSDPA and HSUPA indicate that it is possible to achieve very good bitrates (around 1 Mbps) at large separation distance from the site (over two miles).
- Simulation results can be used as good reference to understand the limits of the planning exercise, however they need to be contrasted with field results.
- Automatic design tools (ACP and DNO) can be very useful to help in the design and optimization of the network, however special care has to be taken when preparing the different inputs to the process.

References

[1] Laiho, J., Wacker, A., and Novosad, T. *Radio Network Planning and Optimisation for UMTS*, John Wiley & Sons Ltd, 2002.
[2] Chevallier, C., Brunner, C., Caravaglia, A., Murray, K., and Baker K., *WCDMA (UMTS) Deployment Handbook: Planning and Optimization Aspects*, John Wiley & Sons Ltd, 2006.
[3] 3GPP Technical Specification 25.101, 'User Equipment (UE) radio transmission and reception (FDD)'.
[4] Qualcomm Engineering Services Group (ESG), 'Aspects of HSUPA Network Planning', Revision A (April 6, 2007) (updated Sept 2008).
[5] Viterbi, A.J., Viterbi, A.M., and Zehavi, E., 'Other-cell interference in cellular power-controlled CDMA'; Communications, IEEE Transactions on, Volume 42, Issue 234, Part 3, FEBRUARY/MARCH/APRIL 1994 Page(s):1501–1504.
[6] Sipila, K., Honkasalo, K.-C., Laiho-Steffens, J., and Wacker, A., 'Estimation of capacity and required transmission power of WCDMA downlink based on a downlink pole equation', Vehicular Technology Conference Proceedings, 2000. VTC 2000-Spring Tokyo. 2000 IEEE 51st Volume 2, 15–18 May 2000 Page(s):1002–1005 vol. 2.
[7] Holma, H. and Toskala, A. *HSDPA/HSUPA for UMTS. High Speed Radio Access for Mobile Communications* (1st Edition), John Wiley & Sons Ltd, 2006.
[8] Pedersen, K.I., Mogensen, P.E., and Kolding, T.E., 'QoS Considerations for HSDPA and Performance Results for Different Services', Vehicular Technology Conference 2006. VTC-2006 Fall. 2006 IEEE 64th Sept. 2006 Page(s):1–5.
[9] Pedersen, K.I., Frederiksen, F., Kolding, T.E., Lootsma, T.F., and Mogensen, P.E., 'Performance of High-Speed Downlink Packet Access in Coexistence With Dedicated Channels', Vehicular Technology, IEEE Transactions on Volume 56, Issue 3, May 2007 Page(s):1262–1271.
[10] Pedersen, K.I. and Michaelsen, P.H. 'Algorithms and Performance Results for Dynamic HSDPA Resource Allocation', Vehicular Technology Conference, 2006. VTC-2006 Fall. 2006 IEEE 64th, Sept. 2006 Page(s):1–5.
[11] Wigard, J., Boussif, M., Madsen, N.H., Brix, M., Corneliussen, S., Laursen, E.A., 'High Speed Uplink Packet Access Evaluation by Dynamic Network Simulations', Personal, Indoor and Mobile Radio Communications, 2006 IEEE 17th International Symposium on Sept. 2006 Page(s):1–5.

[12] Helmersson, K.W., Englund, E., Edvardsson, M., Edholm, C., Parkvall, S., Samuelsson, M., Wang, Y.-P.E., and Jung-Fu, Cheng,'System performance of WCDMA enhanced uplink', Vehicular Technology Conference, 2005. VTC 2005-Spring. 2005 IEEE 61st, Volume 3, 30 May-1 June 2005 Page(s):1427–1431 Vol. 3.

[13] Cozzo, C. and Wang, Y.-P.E. 'Capacity Improvement with Interference Cancellation in the WCDMA Enhanced Uplink', Personal, Indoor and Mobile Radio Communications, 2006 IEEE 17th International Symposium on, Sept. 2006 Page(s):1–5.

[14] Siomina, I. and Di, Yuan 'Enhancing HSDPA Performance Via Automated and Large-Scale Optimization of Radio Base Station Antenna Configuration', Vehicular Technology Conference, 2008. VTC Spring 2008. IEEE, 11–14 May 2008 Page(s):2061–2065.

[15] Lei, Chen and Di, Yuan 'Automated Planning of CPICH Power for Enhancing HSDPA Performance at Cell Edges with Preserved Control of R99 Soft Handover', Communications, 2008. ICC 2008. IEEE International Conference on, 19–23 May 2008 Page(s):2936–2940.

[16] Bertinelli, M. and Malkamaki, E. 'WCDMA enhanced uplink: HSUPA link level performance study'; Personal, Indoor and Mobile Radio Communications, 2005. PIMRC 2005. IEEE 16th International Symposium on, Volume 2, -11–14 Sept. 2005 Page(s):895–899 Vol. 2.

[17] 3G Americas, 'Data Optimization. Coverage enhancements to improve Data Throughput Performance', October 2006.

[18] Wille, V., Pedraza, S., Toril, M., Ferrer, R., and Escobar, J.J., 'Trial results from adaptive hand-over boundary modification in GERAN', IEEE Electronics Letters, Volume 39, Issue 4, 20 Feb 2003 Page(s):405–407.

[19] Toril, M., Pedraza, S., Ferrer, R., and Wille, V., 'Optimization of handover margins in GSM/GPRS networks', Vehicular Technology Conference, 2003. VTC 2003-Spring. The 57th IEEE Semiannual, Volume 1, 22–25 April 2003 Page(s):150–154 vol. 1.

[20] Toril, M., Pedraza, S., Ferrer, R., and Wille, V., 'Optimization of signal level thresholds in mobile networks', Vehicular Technology Conference, 2002. VTC Spring 2002. IEEE 55th, Volume 4, 6–9 May 2002 Page(s): 1655–1659 vol. 4.

[21] http://www.optimi.com/software_applications/x_parameters.

6

HSPA Radio Performance

As has been described in previous chapters, the HSPA and Rel.'99 packet data transmission mechanisms are fundamentally different. The goal of HSPA is to effectively deliver a wide range of different data applications; therefore the associated HSPA functions are all designed to optimally achieve the maximum data capacity over the air interface. With the introduction of important new features, such as shared HS-DSCH, fast packet scheduler, link adaptation and shorter TTI, significant improvements over Rel.'99 for data performance are achieved.

Although operators started to deploy UMTS as early as 2001, it was not until 2005 that the first HSDPA network went on the air. The limited availability of HSDPA capable terminal devices added further delay on the delivery of the technology to the mass market. Most of the existing studies of HSPA performance are based on simulation results from the wireless vendors or research institutions, due to the lack of field test data from real networks. Those simulation results can be good references for operators to understand the performance bounds of the technology. However, they typically do not represent real-world HSPA performance because the assumptions used in the simulation tools generally cannot capture all the limitations and characteristics of the radio conditions of a real network.

The purpose of this chapter is to present early lab and field HSPA trial results to evaluate the HSPA performance under realistic radio conditions. Because most of those tests were completed between 2005 and 2008, when the majority of UTRAN systems deployed were on pre-Rel.'7 platforms, new features such as uplink Discontinuous Transmission (DTX) (gating), Discontinuous Reception (DRX) and F-DPCH were not available for field performance evaluation. The discussion of these trial results reveals the limitations of existing releases and the need for features which improve overall HSPA performance. In addition, realizing that the traffic profile and priority play extremely important roles in optimizing the data network, much attention will be paid to the following: evaluating the performance for different applications; and to optimizing the RAN parameters to best fit the actual packet traffic characteristics of different applications. It is clear that performance optimization of packet data services is exceedingly different from that of traditional voice centric networks. End-to-end performance management for a data network

HSPA Performance and Evolution   Pablo Tapia, Jun Liu, Yasmin Karimli and Martin J. Feuerstein
© 2009 John Wiley & Sons Ltd.

not only spans across different network nodes (as is the case for a traditional circuit-switched network), but also runs vertically through all protocol layers within the same network element.

In the following sections, both lab and field evaluation results are presented to demonstrate the performance of HSPA features under different radio conditions. HSDPA results are evaluated first, then HSUPA. Because all the test systems were pre-Rel.'7, some of these performance results may change after the deployment of HSPA+ in Rel.'7. The feature and functionality of HSPA+ are discussed in detail in Chapter 8.

6.1 HSDPA Lab Performance Evaluation

The main purposes of evaluating HSPA performance, whether HSDPA or HSUPA, in a lab environment before field testing were to validate the basic feature functionality and create a reference for baseline performance of each infrastructure vendor's platform. As is discussed in the subsequent sections of this chapter, many RAN features are heavily vendor-dependant due to the loose definition in the 3GPP standards. Therefore, it is particularly important to establish a lab performance baseline on a per-vendor basis for each feature to be evaluated. Such a rigorous approach helps the operator gain a deeper understanding of the features, as well as identifying during the early stages possible areas for improvement in the vendors' algorithms.

6.1.1 Lab Setup

To avoid interference from other equipment in the lab and have all tests performed under a controlled radio environment the test bed was set up inside a RF shielded room. Figure 6.1 shows

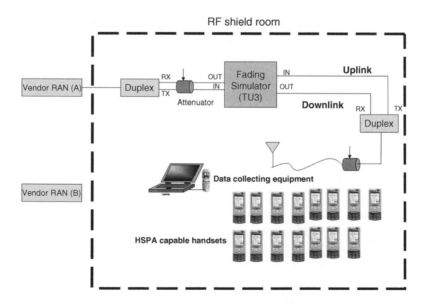

Figure 6.1   Example lab setup for HSPA testing

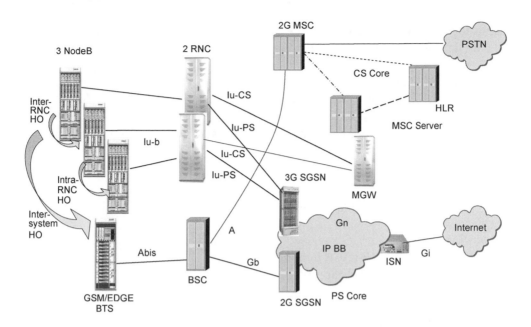

Figure 6.2   Lab trial network diagram

a typical setup for the lab testing. External connections allowed different infrastructure vendor's
RAN equipment to be used. A fading simulator is inserted in the signal path to generate different
radio channel conditions. The radio channel profile in the simulator is remotely configurable
through a computer interface, where the user can either select predefined 3GPP channel models
or define custom channel profiles. Uplink and downlink channels are independently configurable. Duplexers are used to separate the uplinks and downlinks through the simulator. The
pathloss is controlled through adjustable attenuators to emulate different radio conditions.

Figure 6.2 represents a network diagram of the lab network. The network setup allows testing
of most possible mobility scenarios. The lab network includes three NodeBs connected to a pair
of RNCs with both packet-switched (PS) and circuit-switched (CS) core networks. For the PS
core, the 2G and 3G Serving GPRS Support Nodes (SGSNs) were separated in the lab system.
Wireless service providers have the option of combining the 2G and 3G SGSNs in a real
commercial network deployment. For test purposes, an internal FTP server was connected
directly to the Gateway GPRS Support Node (GGSN) to reduce the delay and variability caused
by internet routing.

6.1.2 Basic Functionality Testing

For each infrastructure platform under testing in the lab, baseline performance was established
by performing basic functionality tests in a non-faded environment. Key performance
indicators (KPIs) such as peak throughput, ping latency and call setup time were collected
as the reference for comparison against future test results. These tests typically were run under

three pathloss conditions: good (RSCP > −80 dBm), fair (−80 dBm > RSCP > −95 dBm) and poor (RSCP < −95 dBm). The following are examples of the baseline tests performed:

- maximum cell throughput
  - typically single user with FTP application
- latency
  - round trip time (RTT) for ping tests (32 bytes payload typically used)
- modulation
  - QPSK vs. 16 QAM
- scheduler
  - Round Robin (RR) vs. Proportional Fair Scheduler (PFS)
- PDP context setup time
- channel switching
  - Cell-DCH to/from Cell-FACH to/from Cell-PCH
- Intra-RNC HS-DSCH cell change
- Inter-RNC HS-DSCH cell change
- Inter-Radio Access Technology (IRAT)

All those tests were performed for different vendor combinations to establish a reference point for each vendor platform. Most of these were repeated for different channel conditions after the baseline performance was established.

In the following sections, we focus on features related to the HSPA performance such as latency, mobility, packet scheduler and AMC efficiency.

6.1.3 HSDPA Latency Improvement

The end-to-end network latency includes the contribution from each node on the transport path. To remove the randomness of the internet routing delay, an internal lab application server which was directly connected to the Gateway GPRS Support Node (GGSN) was used to measure the round trip time (RTT) of the network.

A 32 byte ping was used as the standard to measure the RTT value. Figure 6.3 shows the breakdown of the RTT contributing components of a 32 byte ping test, for the condition where the radio bearer was DL HSDPA with UL 64 kbps. From the figure, note that for a total RTT of 92.7 msec, the major contribution was from the Uu radio access link between the UE and UTRAN accounting for 81% of the total delay. The next largest component was from the Iub link between the NodeB and RNC, which contributed 15% of the RTT delay. Together these two factors account for more than 95% of the RTT delay.

Comparing the HSPA RTT with Rel.'99 packet data (typically > 180 ms RTT) or EDGE (typically 300 + ms RTT), the latency improvement from HSDPA ranges from about two to over three times better. However, in HSDPA the Uu air interface still contributes about 80% of the total delay even under the ideal lab condition. In the case of a HS/384K bearer the latency could be reduced by an additional 25 ms.

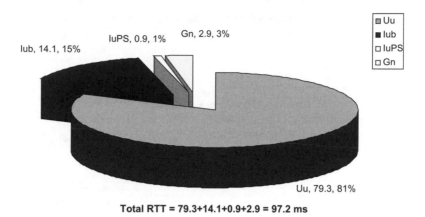

Total RTT = 79.3+14.1+0.9+2.9 = 97.2 ms

Figure 6.3   RTT breakdown of a 32 byte ping test

6.1.4 HSDPA Throughput and Link Performance

Throughput tests in general were used to provide a reference which measures the maximum capability of the system. Obviously, the maximum throughput depended on the UE category. Since most HSDPA devices available in the market typically only supported five codes, the throughput results demonstrated in those tests represent a maximum of five codes only. The UE category also determined the supported modulation schemes. For instance, a Cat 6 device capable of supporting 16 QAM may be able to provide maximum throughput of 3.6 Mbps, but for a Cat 12 device limited to only QPSK, a maximum of 1.8 Mbps is achievable.

Figure 6.4 shows the result of one test example in which a Cat 12 device was used to test the HSDPA user throughput. The average throughput was around 1.5 Mbps which was close to the maximum data rate the system can deliver using five codes and QPSK modulation.

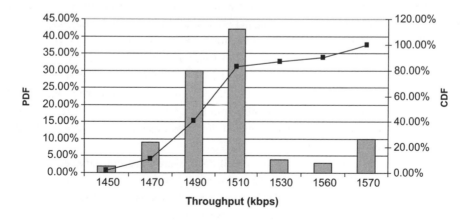

Figure 6.4   HSDPA user throughput in a lab environment (Cat 12 device)

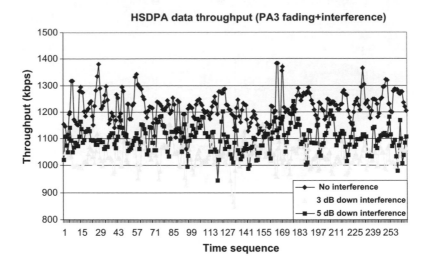

Figure 6.5   HSDPA user throughput under different interference and fading conditions (Cat 12 device)

The throughput was extremely steady as expected because there was no fading or interference involved. To provide a better reference for the field testing, a fading simulator was used in the lab to generate different channel conditions. The downlink neighbor cell interference was created by combining signals from other NodeBs in the lab. Figure 6.5 shows the HSDPA user throughput under various interference levels with a PA3 channel profile. It can be seen from the figure that the neighbor cell interference had a direct impact on the HSDPA cell throughput. An interferer which was 3 dB lower than the signal of the serving cell led to a 15% reduction in the cell throughput. In a field environment, the interference condition will be more complicated. The number of interferers and the distribution of interference will be more random and variable. It is expected that more performance degradation can be observed in the field under high load conditions.

In Chapter 5, a link budget analysis for HSDPA at cell edge was performed based on assumptions for the resource allocation and network conditions. The link budget result can be verified in the lab applying an adjustable attenuator to the radio path (see Fig. 6.6) to simulate the 'drive toward the cell edge' scenario. Both Rel.'99 data and HSDPA were tested for comparison purposes. The results indicated that HSDPA had far better throughput at the cell edge than Rel.'99 data. The Rel.'99 data call dropped at received pilot power values (RSCP) around −115 dBm while HSDPA did not drop until RSCP was below at −120 dBm. Furthermore, the HSDPA throughput was maintained at 800 kbps even in low coverage conditions, until the call eventually dropped.

The reasons for RRC disconnection (data call drop) were different between Rel.'99 and HSDPA: Rel.'99 data call was dropped due to downlink power shortage, while in the case of HSDPA the drop was due to uplink power shortage. Since there was no cell loading and network interference, these results represent the best case scenario in terms of the radio channel conditions.

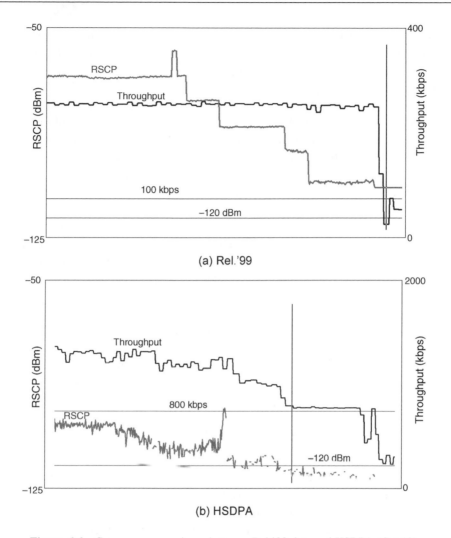

Figure 6.6   Coverage comparisons between Rel.'99 data and HSDPA (Cat 12)

6.1.5 HSDPA Link Adaptation Performance

In HSDPA, adaptive modulation and coding (AMC) is the core algorithm for Link Adaptation (LA). The efficiency of the LA algorithm is critical to the HSDPA performance. Although general requirements for AMC have been defined in the 3GPP standard with respect to the expected performance criteria, the actual implementation is vendor dependent. It is important to understand that the actual performance can vary significantly depending on the vendor. As has been noted, the accuracy of CQI reporting is important for AMC to function properly. However, since the CQI calculation is not clearly defined in the standard, the interpretation of the standard by different handset and chipset vendors can be different. This leads to discrepancies in the CQI reporting among different mobile devices. There have been standards efforts to improve the CQI reporting, but the actual adoption by handset vendors is not mandatory. Therefore it is reasonable to believe that this handset dependency on performance will be present in the market for some time to come.

Considering the vendor dependencies of the LA algorithm, significant efforts were made to establish baseline performance in the lab. Each test case was measured under the same configuration for different RAN platforms and handsets. Since the approach of each vendor's algorithm may be different, and the assumption used by each vendor to optimize the algorithm could vary, it is expected that some vendor's algorithms work better under specific conditions. For instance, an implementation focusing on having maximum throughput will always try to allocate more resources to HSPA and have better throughput performance, while an algorithm weighting more on interference control will be more conservative when assigning system resources; therefore, the throughput tends to be a little bit lower. From an operator's perspective, the algorithm with more focus on the interference control will benefit the whole network and should provide better overall performance in a real environment. Understanding this underlying difference, the engineer will be able to better predict what can be achieved in a real RF environment and which platform could deliver better performance under different network conditions.

As mentioned, CQI reporting is very loosely defined in the 3GPP standards [1], which only indicates a reference measurement period and a mapping between the measurement and the transmitted value for certain predefined performance. The CQI itself is derived from the CPICH quality and later mapped to an estimated HS-DSCH quality. For this purpose the UE assumes that the power in the HS-PDSCH channel is:

$$P_{HS\text{-}PDSCH} = P_{CPICH} + \Gamma + \Delta$$

The measurement power offset Γ is signaled by higher layers, while the reference power adjustment Δ is tabulated based on UE category. The Γ value depends on the vendor's implementation. Since the CQI reporting is not consistent among vendors, a RAN feature called CQI adjustment is implemented on the network side. This feature allows the network to apply an offset to the CQI value dynamically so that the retransmission rate can be kept at a preselected value, for example at 10%. In different implementations, this target retransmission rate could be set differently. The exact implementation depends on how the parameters were optimized and what assumptions were used during the vendor's product development.

Two link adaptation algorithms from different vendors were tested and analyzed. They are referred to as Algorithm1 and Algorithm2 in this section. Algorithm1 has the option of turning CQI adjustment ON or OFF and Algorithm2 has the parameter hard-coded and always has the feature ON. Figure 6.7 shows the performance of the algorithms for different radio conditions (CQI values), represented as the amount of radio transmission failures (NAK rate). As it can be observed, Algorithm2 is more conservative and tries to keep the BLER around 10% [2].

There is a balance between keeping the retransmission rate low by assigning low Transport Block (TB) sizes and being aggressive and having a higher transmission failure rate.

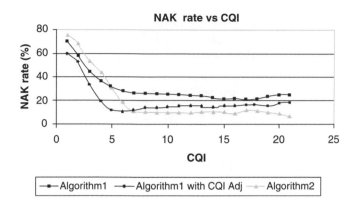

Figure 6.7   NAK rate vs. CQI for different Link Adaptation algorithms

The decision relies on the traffic load distribution and the engineering design objectives of the network. Figure 6.8 represents the effective performance experienced by the user under lab test conditions. It can be seen that an aggressive LA implementation (Algorithm1) leads to better performance towards the edge of the cell. With Algorithm1 the CQI adjustment feature keeps the same throughput level with an overall reduced error rate, which results in a more efficient utilization of the sector resources.

As it can be observed, a proper implementation of the LA algorithm has a significant impact on the end-user's experience. In the case study presented, the user bitrate gains range between 20% and 160% comparing between the LA algorithms, depending on the radio conditions. An aggressive LA algorithm with CQI adjustment provides the best tradeoff between user bitrates and network capacity.

6.1.6 Dynamic Power Allocation

The power allocation in HSDPA is a key algorithm to share the resources between voice and HSDPA data traffic. In Chapter 4 we discussed two types of Power Allocation (DPA) methods

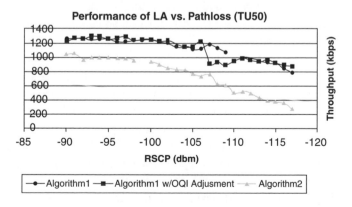

Figure 6.8   Single user throughput vs pathloss for different Link Adaptation algorithms

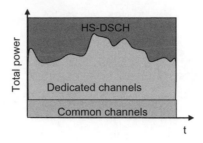

Figure 6.9   HSDPA dynamic power allocation algorithm

for HSDPA: Static and dynamic power allocation. In this chapter, we will compare and contrast the performance achieved by the two allocation schemes.

When static power allocation is applied, a fixed NodeB power is reserved for HSDPA, thereby establishing a cap on the maximum achievable throughput in the sector. This allocation method is not efficient, because even when the cell loading is low, HSDPA still cannot go beyond the reserved power to achieve the maximum available cell data capacity. Therefore, static power allocation is not recommended for HSDPA deployment.

Dynamic power allocation algorithms adjust the HSDPA power based on the current cell loading. Only the unused portion of the total cell power can be assigned for HSDPA traffic. Figure 6.9 shows the scenario when the total 'remaining' power is allocated to HSDPA, which allows maximal HSDPA throughput given particular dedicated channel loading conditions.

In real implementations the operator has the choice of not utilizing the maximum cell power by setting a HSDPA power margin which is defined as the difference between the maximum remaining power and the allowed HSDPA power. To control interference, operators may choose to apply this HSDPA power margin. For instance, for a 20 W carrier configuration with HSDPA traffic, when there is no HSDPA power margin the total cell transmit power will be 43 dBm (20 W). If 1 dB HSDPA margin is applied, the total cell transmit power with HSDPA traffic will be no more than 42 dBm. In a real cluster environment, this parameter should be optimized based on the interference conditions and traffic distribution. As we will see when analyzing the field result, an excessive HSDPA power load can damage the performance of existing voice users in the cell.

The simplest method for DPA is to allocate all the remaining power (considering the HSDPA power margin) at every TTI without considering the TB size and the user's channel condition. This means always having the NodeB transmit the maximum allowed power in all time slots. It is a simple algorithm which takes away one variable for optimization, and the engineer can simply focus on other LA related parameters without worrying about the power assignment of the cell. The disadvantage of this implementation is that the system may transmit more power than needed in certain scenarios; for instance, users near the site with good channel conditions may not need as much power on the HSDPA channel as those in marginal radio coverage, and transmitting at full power to those users will create unnecessary interference to other users in the cluster. In a dense urban environment with high cell densities, interference control is critical. Therefore, for a system running this algorithm, a higher HSDPA power margin may be used to avoid excessive interference introduced by HSDPA.

In a more complicated implementation, the NodeB not only evaluates how much power is available, but also takes into account the TB size and user channel condition and allocates the power accordingly. The idea is to only transmit the minimum power that is needed, i.e. there is additional power control (PC) in addition to the dynamic power allocation. From an interference management point of view, this is a better algorithm. However, it was observed that the reduction of HSDPA power lead to a variation of the Ec/No of the cell which in turn affects the estimation of the CQI value by the users. This creates a feedback loop in the CQI measurements: the network reduces the HSPDA power when the UE reports a good CQI, and in turn this will result in a poorer CQI estimation due to the reduced power offset in the next TTI. This is especially critical when the power fluctuations are large and occur very frequently. Figure 6.10 shows the fluctuation of Ec/No values even when the UE was under a fixed radio condition, for a continuous FTP download.

Since a fading simulator was not used in this case and only HSDPA traffic was served by the cell, the fluctuation of the Ec/No values reflects the transmit power variation created by the CQI feedback loop noted above. Since 16QAM has larger TB sizes, more power was assigned (lower Ec/No as measured by the mobile). As it can be observed, the power fluctuations are in the order of 2 dB.

This CQI feedback loop effect, together with the reduced amount of power allocated for HSDPA in some cases has been shown to greatly degrade the performance compared to a simpler DPA technique. The performance degradation was remarkable in low load environments (36% loss), but was especially detrimental in high load cases, reaching up to 78% loss

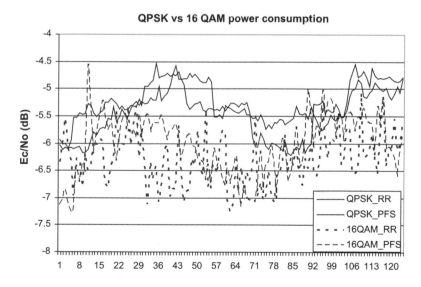

Figure 6.10 DPA power control for different modulation schemes (QPSK and 16QAM) and packet scheduler algorithms (RR = Round Robin, PFS = Proportional Fair Scheduler)

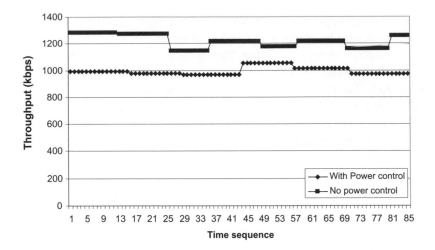

Figure 6.11 Dynamic power allocation implementation comparison (single cell with 40% loading). Single user throughput (Cat 12 device)

with 60% voice traffic load. As shown in Figure 6.11, for a single cell with 40% real-time traffic loading, the simple algorithm (i.e. no PC) shows better performance than the one with PC scheme. This was expected since the simple algorithm always transmits at the maximum allowed power level while the power control algorithm tries to adjust the transmit power level of HSDPA.

Later in Section 6.3.2.3, we demonstrate that the field performance of the DPA with power control results in a significantly degraded HSDPA capacity as compared to the full power assignment, especially when voice services share the same power resources.

6.1.7 HSDPA Scheduler Performance

Three scheduler algorithms are supported by most RAN platforms: Round Robin (RR), Proportional Fair Scheduler (PFS) and Max C/I. All three were investigated under different fading environment simulated by the fading simulator. Table 6.1 summarizes the test results. The notable gain of PFS and Max C/I over RR could only be observed for low mobility scenarios. Under high mobility (Typical Urban 50 km/h or TU50), there was no significant differences among the three schedulers. In a stationary case, PFS gain over RR was negligible. For Max C/I, the gain over RR will vary depending on the UE relative conditions.

In a practical environment, we can expect that most mobile applications have small object sizes to download, such as HTML web pages when internet browsing. Therefore, latency performance will be the number one priority for achieving a better user experience. Since the Max C/I algorithm always favors users in better radio condition, it runs the risk of starving users under poor radio condition. Therefore, although the overall cell throughput could be better when Max C/I scheduler is applied, it is not recommended for a network with many active data

Table 6.1 HSDPA scheduler relative performance under different channel conditions

Fading Profile	Average PFS gain for cell (across nominal CQI range 1–13)	Average Max C/I gain for cell (across nominal CQI range 1–13)
Pedestrian-A 3 kmh (2 users)	+38%	+25%
Vehicular-A 3 kmh (2 users)		+16%
Typical Urban 3 kmh (2 users)		+12%
Typical Urban 50 kmh (2 users)	+5%	+4%
Stationary Over the air (3 users)	+0%	variable

users. PFS, on the other hand, considers each user's radio condition and the fairness of the capacity allocated to each user. It is the algorithm we recommend for the deployment.

Section 6.3.2 presents the performance of PFS in a field environment, in which it can be seen how the users in worse radio conditions experience significant improvement in the performance without sacrificing other users. In general PFS is an effective way of ensuring a homogeneous user experience across all users in the sector, which is why we recommend it for operators.

6.2 HSUPA Lab Performance Evaluation

Fundamentally, HSUPA has more similarity to Rel.'99 than HSDPA due to its dedicated nature for physical channel assignment. Less coding protection and lower spreading gain for HSUPA implies more power consumption per data link. This leads to increased uplink interference and potential performance degradation. Therefore, in our HSUPA studies, we placed great emphasis on link budgets, uplink capacity and interference analysis.

6.2.1 Throughput Performance

Figure 6.12 shows the individual user throughput under different pathloss conditions without fading. The results showed that maximum user throughput was reached at 1 Mbps for a Cat 5

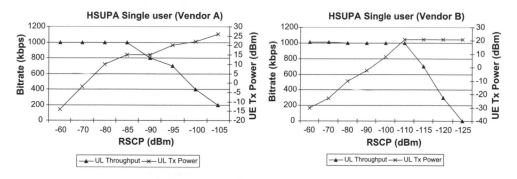

Figure 6.12 Single HSUPA user UL throughput and transmit power performance in different vendor implementations (Vendor A vs Vendor B)

HSUPA device and an almost linear degradation of the throughput until the UE reached the cell edge. As can be observed, there was a major difference in the cell edge performance depending on the vendor implementation. In the case of Vendor A the coverage footprint is about 5 dB lower than the one observed for Rel.'99 data (64 kbps), and about 15 dB lower than the one observed in System B. As mentioned in Chapter 5, it is important to revisit the link budget assumptions after having analyzed the performance of the real network for each particular vendor implementation.

Another important observation from Figure 6.12 is the high values of the UE transmit power required to support the HSUPA operation, which is between 10 and 15 dB higher than the reference Rel.'99 64 kbps channel requirements. This represents a large power difference for a mobile device and illustrates that the increased uplink data rates of HSUPA come at the cost of additional power consumption.

Figure 6.13 shows the increased cell throughput for Vendor A when two users transmitted simultaneously: up to 1.6 Mbps was achieved with two simultaneous users, with an average of 800 kbps per user.

6.2.2 Scheduler Performance

One major difference between HSUPA and HSDPA schedulers is that HSUPA mainly uses power control to overcome the channel fading and manage the uplink interference at the same time. The 'interference budget' for each HSUPA user is controlled by the HSUPA scheduler in the NodeB which overlooks the received total wideband power (RTWP) on the uplink and determines if a power up request (called the happy bit, as noted earlier) from a particular UE can be granted. To evaluate the efficiency of the scheduler, the tests were repeated under fading conditions. Figure 6.14 shows the uplink cell throughput with a pedestrian (PED-A) channel profile. In this test, there were two FTP users under similar radio conditions. The idea is to understand how the scheduler assigns priorities to different HSUPA users. As can be seen, the aggregate cell throughput had a reduction of 15% as compared to the 'ideal' case with no fading.

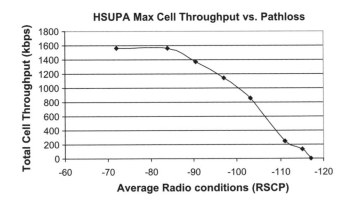

Figure 6.13   HSUPA cell throughput for two users without fading for Vendor A implementation

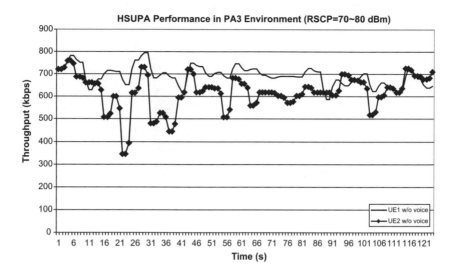

Figure 6.14 HSUPA user throughput under PED_A channel profile, two HSUPA users in the cell with no voice traffic

Figure 6.15 illustrates the performance of the scheduler when one of the two users moves toward the cell-edge coverage, its performance is greatly degraded as compared to the other user. The system is not performing a fair distribution of throughputs.

We observed that while infrastructure vendors have devoted much effort to improving their HSDPA scheduler algorithms, the HSUPA schedulers still need further improvement. The

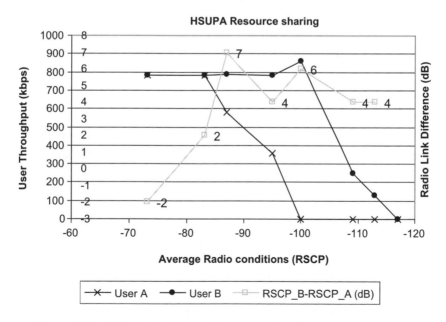

Figure 6.15 HSUPA scheduler performance under different radio conditions

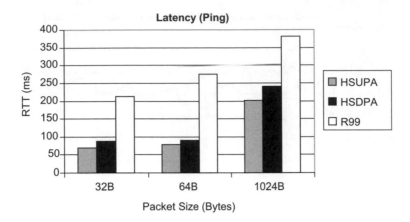

Figure 6.16 HSPA latency improvements

overall performance could be improved employing by smarter methods, such as applying the Proportional Fair concept to the distribution of transmit power rather than just scheduling times.

6.2.3 Latency Performance

In addition to the maximum throughput improvement, another major benefit of utilizing HSUPA is the latency reduction. As it is indicated on Figure 6.16, HSPA has 20% improvement over HSDPA/R'99 when HSUPA uses the 10 ms TTI frame (further gains are expected from the 2 ms TTI frame). Comparing pure Rel.'99 to HSPA, the latency reduction is around 200%. Therefore HSUPA will not only benefit applications which need bigger bandwidth, but also those with small objects such as web browsing, instant messaging etc.

6.2.4 Mixed Voice and HSUPA Performance

Data traffic has different characteristics compared to voice. Data transmission characteristics are highly dependent on the application type and content structure. When radio resources are shared between these two types of traffic (for instance, a single carrier with both voice and data support), the radio resource management (RRM) algorithm should be assigning the highest priority to the voice service. A test case was designed to have both voice and HSUPA traffic activated on the same carrier under two different radio conditions (received pilot power at -80 dBm and -95 dBm respectively). The voice traffic introduced (15 simultaneous voice users) represented about 30% of the designed voice capacity. As can be observed in Figure 6.17 the HSUPA throughput was significantly reduced when the voice traffic was present. In this case, the throughput degradation was about 100% under good radio conditions (no neighbor interference, RSCP ~ -80 dBm). Voice call drops where observed when the HSUPA call was added to the cell.

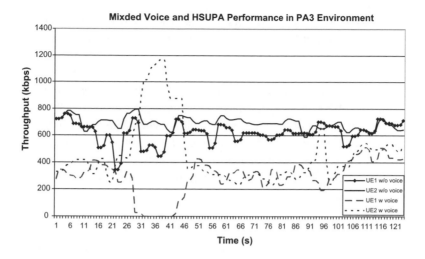

Figure 6.17 Voice traffic impact on HSUPA throughput

In degraded radio conditions, the link budget shortage will force UEs to power up and compete for limited resources. Figure 6.18 shows HSUPA throughput reduction was significant (down to below 200 kbps) for the case when UEs were in conditions with poor radio signals (RSCP ~ -95 dBm). One of the data calls was dropped during the process, and other voice

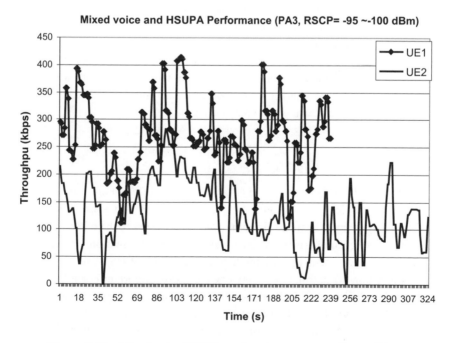

Figure 6.18 Mixed voice/HSUPA performance at poor radio conditions

calls dropped as well. At this point, the system was reaching overload thresholds for uplink received power (noise rise); therefore, new voice call originations were blocked due to lack of radio resources.

The results showed the substantial impact of HSUPA on voice quality as well as on the uplink capacity. In a typical CDMA system, downlink is normally the capacity bottleneck. However, the introduction of HSUPA on the uplink may present noticeable changes to the noise rise at the NodeB receivers. Since this test was done in a lab setting where UEs are close to each other, the likelihood of having correlated radio path among UEs was high. Therefore, the combined uplink noise from all UEs would be higher than in a field environment where UEs can be distributed across different locations within the cell coverage area. UEs in field conditions will typically have more independent channel characteristics from each other compared to these lab scenarios.

In Section 6.3.3, we will present field results on HSUPA, where the severe degradation noted in Figure 6.18 did not occur. That highlights the importance of performing a two-step evaluation of the technology, based first on lab test evaluations and later on actual field measurements. Although lab testing is a very useful process to identify future network challenges, these tests alone cannot capture the full extent of conditions observed in field situations.

6.3 Field Evaluation

There are several areas that cannot be effectively tested in a lab environment, including the following: mobility and performance with cluster interference. In addition, HSPA capacity and its impact on voice performance need to be evaluated in the field since the results collected in the lab cannot fully represent real network conditions. To completely evaluate the HSPA performance, it is important to have all features tested under a real network environment. For a field system, depending on the operator's legacy system, the configuration varies from one to the other, or even from one market to the other. It is important to categorize the RF environments so that the test results are comparable.

6.3.1 Field Network Configurations

Field results presented in this section were collected in five different clusters which represent typical network topologies for most operators, as described in Table 6.2. Since the network configuration and vendor platform are different from one network to the other, it is not practical to have a side-by-side comparison for each feature under evaluation. However, from an operator's point of view, understanding the field performance of HSPA under different radio environments and how to optimize the network are more important considerations. As we continue analyzing the results in this section, our main focus will be on how to design and optimize the HSPA network based on what was learned from the field results.

Mobility, network loading and neighbor interference are the factors which cannot be easily created in lab systems. Therefore in the field, focus was placed on investigations of HSPA

Table 6.2   Field clusters for HSPA feature evaluation

Cluster	Radio Environment	Cluster size	Feature release	DPA
A	Suburban/rural	11 sites	Rel.'4, Rel.'5	Static
B	Suburban	29 sites	Rel.'5	DPA with Power Control
C	Dense urban	35 sites	Rel.'5	DPA with Power Control
D	Urban	19 sites	Rel.'5	Aggressive DPA
E	Suburban	29 sites	Rel.'6	DPA with Power Control

performance under different mobility scenarios and interference conditions. To evaluate the Inter-RNC mobility performance, most clusters were configured to have two RNCs hosting the cluster. On the backhaul capacity side, most of the sites in those clusters have one T1 Iub bandwidth (1.54 Mbps nominal, 1.2 Mbps effective). Considering the peak data rate offered by the network, the Iub links for the test sites were under dimensioned. However, as the discussion of the test results continue, it will become clear that using the peak rate of HSPA to dimension the backhaul capacity is not an effective way for network capacity planning. Both the data traffic models and the air interface limitations will play important roles in this engineering dimensioning process.

All field clusters have one UMTS carrier frequency. The total traffic power is set at 20 W with CPICH power at $10 \sim 15\%$. Since there were no commercial services in these clusters, the traffic loading was generated by using Orthogonal Channel Noise Simulator (OCNS) on the downlink. OCNS is a standard feature which has been implemented in most WCDMA vendors' systems. For HSUPA testing, users compete for the uplink power resource (noise rise), therefore load can be generated by introducing more HSUPA users into the system.

Network geometry, which describes the relationships between the serving and interfering sectors, has direct impact on the radio performance. The characteristics of the interference distribution changes from cluster to cluster. In order to fully investigate the efficiency of the design and algorithms of the HSPA system, many tests were done in both macro and microcell radio environments.

The development of appropriate test plans for field feature evaluation is very important. The ultimate goal is to find the best solution to serve the customer in an efficient way, so the approaches of designing the test cases are rather different from those of research and development (R&D). Using HS-DSCH cell change as an example, a R&D type of test will focus on the success rate and how fast the transitions are without considering the actual user's possible behavior. In reality, a data user is very likely to stay stationary inside the soft handover (SHO) zone where multiple pilot channels can been seen by the mobile devices. Load fluctuation and random channel fading will lead to constant HSDPA serving cell changes for UEs staying in the SHO zone. Although the HS-DSCH cell change procedure can be fast, the constant cell changes which involve a significant amount of RRC signaling and radio channel reconfigurations will cause performance degradation and high interference to the network.

For an operator, understanding the traffic characteristics is the starting point for new feature evaluation, planning and rollout. Below are examples of the HSPA related features that could be evaluated in the field:

- Modulation scheme. QPSK/16 QAM.
- HSDPA power allocation: dynamic vs. static.
- CQI adjustment.
- HSDPA mobility.
- HSDPA scheduler (Round Robin, PFS and maximum C/I).
- HSDPA code multiplexing.
- Rx diversity.
- Interference Cancellation (IC).
- E-DCH resource scheduling.

6.3.2 HSDPA Performance

6.3.2.1 HSDPA Throughput

Although some of the test clusters supported both QPSK and 16QAM at the time of the testing, most sites only had one T1 backhaul capacity, and therefore the benefit of using 16QAM could not effectively be realized in some of the drive tests. Based on our lab results and the CQI distributions observed in the networks (between 40%–60% of the areas with CQI > 16) we can expect significant throughput improvement as compared to the results shown in this section, due to the use of 16QAM instead of QPSK. The expected user throughput and capacity can be calculated considering the DTX factor on the actual measured throughput.

Figure 6.19 shows that 80% of the data points have 800+ kbps throughputs, as collected by drive tests in network A. As described above, only QPSK modulation was configured, which limited the maximum throughput. The average vehicle speed was around 30 miles/h.

It should be noted that there was no traffic loading on the network during the above tests. The drive route was concentrated on the cluster center where coverage was relatively good. Figure 6.20 shows a similar drive test in network C on a different RAN vendor platform demonstrating a similar throughput distribution. The site density in network C is much higher than for network A, with most sites in network C spaced less than 0.5 miles apart. Since the test was conducted without traffic load on the network, system resource and network interference constraints did not cause limitations on HSDPA performance. The data throughout was above 1100 kbs for 80% of the time.

The distribution of throughputs when multiple users transmit at the same time is quite homogenous thanks to the effects of the PFS, as can be observed in the following results from Network D. Figure 6.21 presents the throughput distribution of two users performing continuous FTP downloads with different radio conditions comparing a PFS algorithm vs. RR. User 1 is in a good location, with RSCP levels around −70 dBm, Ec/No around −6 dB and CQI around 16. User 2, on the other hand, is at the edge of the cell, with Ec/No around

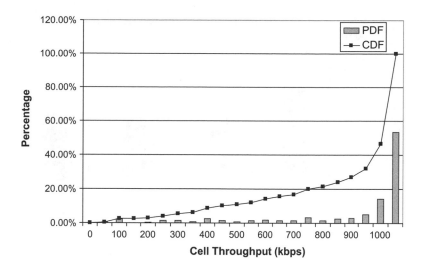

Figure 6.19   HSDPA drive test throughput in cluster A (QPSK only)

−105 dBm, Ec/No around −10 dB and reported CQI around 11. The results indicate that the PFS scheduler is able to increase the throughput of the user at the cell edge (around 20% increase) while not degrading the experience of the user in better conditions, as compared to a RR scheduling scheme.

When both users are moving their average throughputs further equalized. In fact, the average CQI in their respective routes was very similar (CQI 13 for User 1 and 14 for User 2).

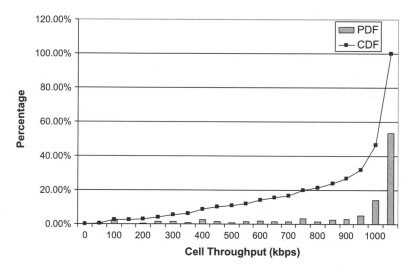

Figure 6.20   Drive test throughput in cluster C (Dense Urban) (QPSK only)

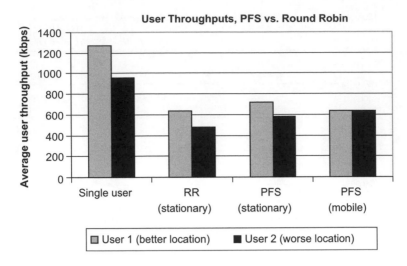

Figure 6.21 Example of throughput distribution with Proportional Fair Scheduler

6.3.2.2 HSDPA Link Budget

In the field, the HSDPA link budget can be verified by driving a test UE from the center of the coverage area to the cell edge. To avoid HSDPA cell changes during the drive test, the selected serving cell should be facing outside toward the UMTS cluster boundary. Since cells on the outskirts of the UMTS cluster are typically less impacted by the neighbor cell interference, the conclusion drawn from these tests may be on the optimistic side. The engineer can use these values as a reference to get the actual link budget by applying the cluster loading factor to the calculation.

Figure 6.22 shows the average throughput of a drive test with one HSDPA UE moving from the cell center to the edge. The test was done under unloaded conditions for two different types of HSDPA allocation methods: a conservative DPA that uses power control (top chart) and an aggressive DPA (bottom chart). The aggressive DPA assigns all the remaining power to HSDPA and as a result higher bitrates can be achieved (between 20%–35% higher throughputs) in good coverage areas, however the results in marginal coverage areas are similar.

The test results show solid HSDPA link performance in unloaded conditions. This test scenario represents a data-only carrier scenario where all cell resources are allocated to HSDPA. In addition, the channel orthogonality was good since neighbor interference was limited for cells at the boundary and there was no loading on any of the cells in the cluster at the time of the testing. Figure 6.23 shows throughput at various RSCP levels based on drive test data in a cluster that is 60% loaded with OCNS. Note that these results were obtained for a network with a conservative DPA scheme (e.g. DPA with power control).

It can be seen that the data performance at the cell edge are most affected when the network is loaded. The throughput reduction is about 30% in the −105 to −95 dBm RSCP range. Analyzed from a link budget perspective, the pathloss to support a data rate of 300 kbps is

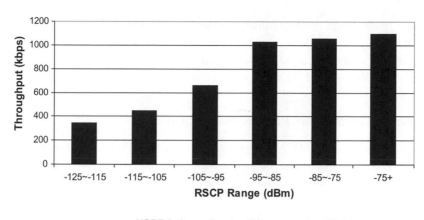

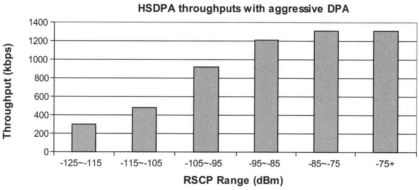

Figure 6.22 HSDPA throughput performance vs. coverage (unloaded) for two different HSDPA power allocation methods: DPA with power control (top) and DPA with full power assignment (bottom)

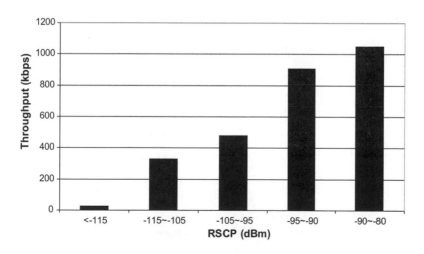

Figure 6.23 HSDPA throughput performance vs. coverage (60% loading)

reduced by 10 dB, from 155 dB to 145 dB. On the other hand, the impact of the network loading to the data throughput inside the good coverage area was not significant.

In summary, the cell coverage for data is much smaller under loaded conditions, however the performance of the users in the good coverage areas will not suffer significantly.

6.3.2.3 Single Carrier Voice Plus HSDPA Performance

After the introduction of HSDPA many operators have raised the following critical question: can voice and HSDPA share the same carrier without degrading each other's performance? It is not a simple yes or no answer because the operators' design coverage objectives vary from network to network.

Considering the expected light load of the UMTS traffic at the beginning of a new deployment, the operators may opt for designs that have lower loading criteria and thus less site density to save on the initial capital spending. For network under light load design goals, the operators should be cautious when introducing HSPA services on top of the initial designed network. The power margin left for HSDPA will have to be reduced to maintain the designed coverage goal even though the HSDPA dynamic power allocation algorithm is capable of assigning the remaining power to HSDPA users. Throughout this book, we have emphasized the importance of understanding the relationship between the design goals for coverage and network capacity in a CDMA network. The tradeoffs between coverage and capacity will be examined further while answering the question of joint optimization for voice and data services on a shared carrier frequency.

Figure 6.24 shows the study of the relationship between the network design goal and the available HSDPA resource. In the RRM function, voice is prioritized over data on the same carrier so that the data service does not cause a degradation to voice performance. The test case in the figure was collected from a network which implements a conservative DPA with power

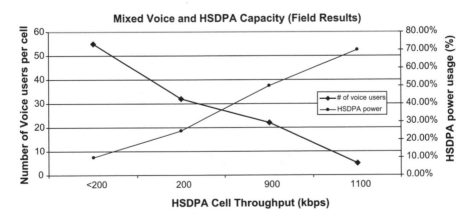

Figure 6.24   Voice and HSDPA capacity sharing on single carrier (DPA with Power control, QPSK only)

control. Per carrier power is at 20 W and the coverage is designed for a fully loaded network, i.e. all power can be allocated for traffic.

We can see that since voice and HSDPA share the same radio resource, increasing the capacity of one will decrease the capacity of the other. With 22 simultaneous voice users, HSDPA demonstrated 900 kbps cell throughout in cluster B where those results were collected without OCNS loading on neighboring cells. The power allocated to HSDPA was around 30% of the total power (6 W). In the early stages of the deployment, as it is indicated in Figure 6.24, the impact of increased voice traffic on HSDPA throughput will not be significant: when the number of voice calls increased from 0 to 22, the HSDPA throughput only reduced by 18% (from 1.1 Mbps to 900 kbps). However, after adding 10 more voice calls, the HSDPA cell throughput decreased to 200 kbps, further reduced by 78%. Obviously the loading curve approaches the capacity limit much more quickly when the traffic reaches critical levels. This has been a well known effect for a CDMA system, and is an important justification for the deployment of data-only carriers, once the voice capacity reaches threshold values that would involve severe tradeoffs between voice and data capacity. Chapter 7 provides more details on carrier deployment strategies.

As mentioned earlier, the implementation of the Dynamic Power Allocation method has a major effect on the expected HSDPA cell throughputs, especially in the presence of voice traffic. As presented in Chapter 4, and further commented on in Section 6.1.6, there are two typical methods for DPA, a more aggressive one in which all the remaining sector power is allocated to the HSDPA service and a more conservative approach that applies a certain type of power control over the available HSDPA power. In a cluster with a more aggressive DPA between 40% and 80% higher HSDPA capacity could be achieved (see Figure 6.25). Tests in indoor locations with inbuilding sites have also shown higher HSDPA and voice capacities, with up to 50 voice calls plus 2 Mbps in a single sector.

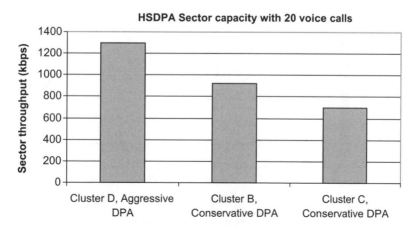

Figure 6.25 HSDPA+Voice capacity depending on DPA scheme (no OCNS) illustrating throughput improvement with aggressive DPA scheme

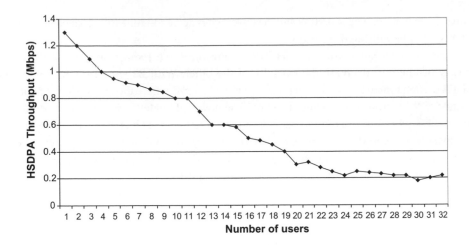

Figure 6.26 Voice and HSDPA capacity sharing on single carrier with cluster OCNS loading at 60% (DPA with Power Control)

When OCNS is used to load the neighboring cells, the saturation point will be reached even earlier as shown in Figure 6.26. It is obvious that HSDPA performance will suffer when the overall network load reaches a critical level where in this case, it is 60% loading or 4 dB downlink noise rise on the whole cluster. This is one of the major reasons to deploy a HSPA data-only carrier in networks where the overall interference level is high. More detailed will be discussed in Chapter 7.

Since dynamic power allocation allows HSDPA users to use the remaining power of the carrier, the concern over voice call performance needs to be addressed. Figure 6.27 shows the

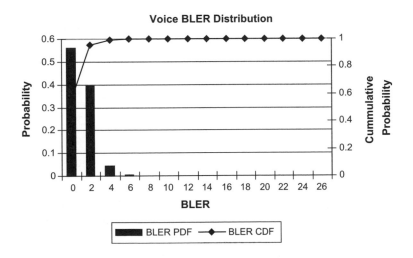

Figure 6.27 Voice call BLER with Mixed Voice and HSDPA traffic test in Cluster A

voice call Block Error Rate (BLER) collected with 22 simultaneous voice calls and two HSDPA data sessions served by one cell in cluster A when the HSDPA power was limited to 6W.

The voice quality degradation is insignificant as shown in this test. In 95% of the data points, BLER values are less than 2%. Some of the high BLER data points come from the areas where coverage is marginal due to the small coverage footprint of the cluster. These BLER values are comparable to the ones obtained in the networks that applied DPA with power control. On the other hand, results from Network D where the HSDPA power allocation was more aggressive indicated higher impact on voice traffic, with only 90% of the BLER samples below 2%. In fact, there were 5% of the samples with BLER $> 4\%$ which is considered a typical threshold for acceptable voice quality. This type of DPA strategy requires some planning effort to determine what should be the right value for the power margin, which depends on the expected voice capacity in the shared carrier.

In summary, the key to operating a single carrier supporting both voice and HSDPA is to set the design goal clearly at the beginning. What will be the expected voice traffic volume? What will be the penetration rate for HSDPA capable handsets, what kinds of data applications are expected, and what is the long term spectrum strategy? The operators should always ask these questions at the beginning of the UMTS network deployment. Good strategic planning leads to solid network design and less stress on the network operation and optimization in the future.

6.3.2.4 HSDPA Mobility

The performance of HSDPA mobility is one of the major focus points for the field testing due to the limitations of the lab test environment. There are three test scenarios in terms of the HSDPA mobility: Intra-Node-B, Inter-Node-B cell change and Inter-RNC mobility. In the control plane there will be no outage during the cell change procedure, since the A-DCH employs soft-handover; however, the outage in the user plane can be longer due to factors such as buffer clearing, retransmissions or TCP slow start during the serving cell change. In addition, the radio conditions play a big part in the performance of the HSDPA cell change. Therefore, the performance of the HSDPA cell change could vary from one area to the other, as is observed from the results collected from different clusters.

Intra-Node-B HSDPA Cell Change
Intra-Node-B HS-DSCH cell change takes place among cells within the same Node-B. In these tests, an FTP download is used to keep continuous data transferring on the HS-DSCH.

In the soft-handover (SHO) zone, both the current serving cell and the target cell pilots will be strong. Since only A-DCH can be in soft-handover and the shared channels (HS-DSCCH and HS-DSCH) can only be established on the HSDPA serving cell, those non-serving pilots in active set will generate interference to the shared channels. This gap could be reduced when the cell coverage overlap is small. Figure 6.28 shows the results collected in cluster D. The network was not loaded during the test. The change in cell scrambling code number indicates the moment of the serving cell change.

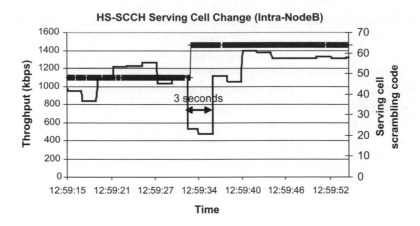

Figure 6.28   Data throughput for HS-DSCH intra Node-B cell change in Cluster D without network load

Test results in different clusters have been consistent showing those low throughput gaps during the intra-Node-B HS-DSCH cell change. The typical range is 2 to 5 seconds depending on the network conditions. Since intra-Node-B HSDPA cell change happens within the same Node-B coverage area, in most of the cases, the number of strong cells is smaller compared to the inter-Node-B scenario.

Inter-Node-B HSDPA cell change (Intra-RNC)
In the Inter-Node-B cell change the MAC-hs buffer from the previous sector cannot be reutilized in the target sector, and this will typically imply a retransmission of data previously sent to the mobile that couldn't be properly acknowledged. This results in a performance degradation compared to the Intra-Node-B case. Furthermore, since at the Node-B boundary, it is more likely that multiple cells have overlapped footprint, the interference condition is more complicated and could result in further degradation, especially in areas with pilot pollution. Figure 6.29 shows that the inter Node-B cell change test in Cluster D has a throughput degradation for almost 6 seconds.

In this case, the HSDPA throughput during the cell change was reduced from 1.3 Mbps to around 500 to 800 kbps. The speed of the drive test vehicle was around 30 miles/hr. This does not necessarily represent the mobility profile for the majority of data users which in general have low mobility or even are stationary. A long dwell time of users in the soft-handover zone will cause degradation of the HSDPA performance as it is indicated in Figure 6.30. In the cell boundary area, before the cell change takes place the throughput is degraded to an average of 600–800 kbps.

For stationary users staying in the SHO zone, the probability that those UEs are experiencing constant cell changes could be much higher due to the channel fading and interference fluctuation caused by the loading change in the cell. The HSDPA data performance for those users will be far below the expected design goals. The operators could raise the hysteresis value to delay the cell change and reduce the ping-pong effect. However, this

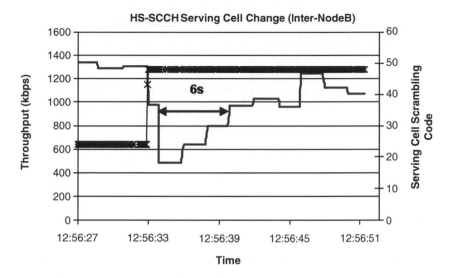

Figure 6.29   Data throughput for HS-DSCH inter Node-B cell change in Cluster D without network load

solution itself doesn't change the interference condition in the SHO zone. Delaying the
HSDPA cell change too much will run the risk of dropping the data call which in turn
degrades the data performance.

In networks with high SHO overhead, more users will have degraded HSDPA service due to
constant HSDPA cell changes. For instance, if the SHO overhead is 1.5, 50% of the total
footprint will be in the SHO zone. It may not present a big problem on voice performance other

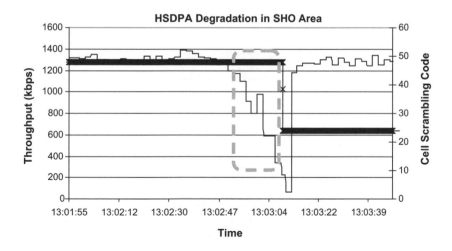

Figure 6.30   Data throughput for HS-DSCH inter Node-B cell change for low mobility use

than consuming more network radio resources, however the impact of overlapped cell coverage on HSDPA users could be substantial. For the operator, keeping the SHO overhead at an optimum level will not only benefit the data performance, it will also provide more Node-B capacity for all users. In general, the recommended SHO overhead should be less than 40%.

In dense urban areas where the site density is high, maintaining low SHO overhead could be challenging. More attention should be paid to the cell power containment through different design approaches. For instance, using outdoor cells to provide indoor coverage runs the risk of generating excessive interference to the outdoor environment. One way to mitigate this problem is to use low power indoor pico cells to provide better network interference management. Other methods could include using distributed antenna systems (DAS) or applying different frequencies for voice and data. As we move on to the capacity growth planning in Chapter 7, the carrier allocation for voice and data will be discussed further.

Inter-RNC HSDPA Mobility

Inter-RNC mobility for HSDPA requires extra signaling involved in the cell change process. This will cause more delay for HSDPA cell change and the low throughput gap is expected to be longer compared to the Intra-RNC mobility cases. The implementation of the inter-RNC HSDPA cell change also varies among different infrastructure vendor platforms. Figure 6.31 shows the case when Serving RNC (SRNC) Relocation is not supported. The Directed Signaling Connection Re-establishment (DSCR) procedure is applied instead in this case. The DSCR procedure releases the connection with the 'DSCR' cause code, which orders the UE to immediately re-establish the connection in the appropriate cell. As we can see the interruption period can be as high as 8 seconds for a vehicle traveling at 30 mph.

In the case of stationary users located inside the RNC boundary, the data performance degradation could be much worse. In the field we observed that many of those users eventually were downgraded to Rel.'99 radio bearers. This procedure could be improved by adding intelligence in the mobility management of the RNC to make sure no DSCR is issued during the active data transferring period, however, the benefit could only be applied to certain applications

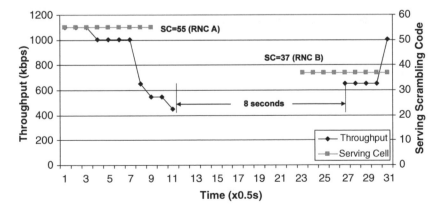

Figure 6.31 Inter RNC HSDPA cell change without SRNC relocation

with pauses during active transfers, such as web browsing. The data performance for stationary users in the transition zone will still be an issue due to the presence of multiple strong pilots (interference) from different RNCs. Fundamentally this is the same interference situation that was faced with the intra-RNC HSDPA, inter-Node-B cell change.

For systems supporting SRNC relocation, in theory the outage time could be shortened since there are no HSDPA release and reestablishment procedures involved. However, the actual field performance will depend on the specific algorithm implementation of the platform and the RF environment. From the field test results collected in different clusters, this is the area that shows the most variation. Figure 6.32 is the result collected during one drive test in cluster D where SRNC relocation is supported. In the figure, there were several occasions when HSDPA was downgraded to Rel.'99. The time span during which the UE stayed on Rel.'99 ranged from 40 seconds to 10 minutes before switching back to HSDPA. In this particular case, the SRNC relocation is not performing up to the operator's performance expectations.

In summary, the mobility performance for HSDPA faces significant challenges in SHO areas where no dominant server is available. Constant measurement reporting events 1d-HS triggered by the fluctuation of the pilot signal strength and the network interference levels causes excessive HSDPA cell change procedures being executed in the SHO zone. This ping-pong effect could be alleviated by applying a penalty timer or increasing the hysteresis, however, these methods don't remove the presence of the strong interfering signals from non-serving cells. Preventing the cell change by using a longer triggering timer will make the UE suffer more from the interference. This fundamental difference between HSDPA and voice requires the design and optimization of each service being treated differently. For certain networks with relatively high soft-handover overhead, a HSPA data-only carrier can be more

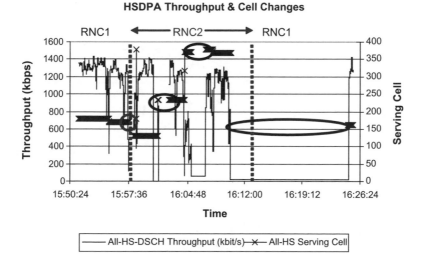

Figure 6.32   Inter RNC HSDPA mobility drive test in cluster D

attractive than mixing voice and HSPA on a single carrier since the data carrier can be designed differently (such as hotspot application).

6.3.3 HSUPA Performance

HSUPA provides improved uplink data performance by introducing Node-B based scheduling and acknowledgment schemes with shorter TTI. From a radio resource management perspective, HSUPA is similar to Rel.'99 since each HSUPA user occupies a dedicated channel and the system still uses fast power control. In HSUPA users are competing for the overall uplink noise budget allowed by the network. Lab results indicated that HSUPA traffic may cause voice call drops and uplink capacity could decrease significantly due to excessive uplink noise rise; however, the field results show a lower impact from HSUPA to voice users' performance. The lab results could be caused by the fact that all test mobiles had to be placed close to each other in the testing room, and therefore the likelihood of having correlated radio path among mobiles was very high. In the field, the noise rise was much lower than the one being observed in the lab.

The field results were collected in cluster E, which is a suburban type of environment. The HSUPA device was Category 5, however the maximum data rate permitted by the network was 1.4 Mbps due to baseband resource limitations.

6.3.3.1 HSUPA Link Level Performance

The HSUPA link performance was validated under different mobility scenarios. Figure 6.33 shows that the average throughput was above 300 kbps during the entire drive test, with peak rates around 900 kbps. This is a single user case and in a medium mobility scenario (35 mph).

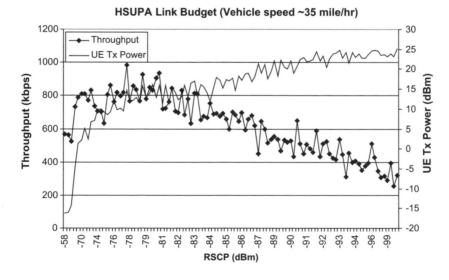

Figure 6.33   HSUPA link budget validation at medium mobility (<35 miles/hr)

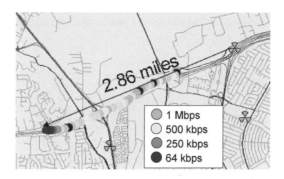

Figure 6.34   HSUPA link budget validation (unload at 60 miles/hr)

The UE transmit power was higher than 10 dBm most of the time (even in very good radio conditions). For cells supporting multiple HSUPA users, the high UE Tx power could be a potential interference issue, especially for cells with small coverage and high traffic density.

Figure 6.34 shows the drive route with throughput data for the high mobility (60 mph) test scenario. HSUPA throughput was above 600 kbps even when the UE was 2 miles away from the cell tower. The HSUPA call dropped when the mobile was 2.8 miles away from the cell due to loss of coverage.

Figure 6.35 shows that the UE transmit power was at 10 dBm most of the time and lower than the medium mobility case. This is the same drive route as above.

The results of the drive tests are in line with the expected throughput vs. pathloss from the lab results. As we mentioned in that section, the vendor implementation has a significant impact on

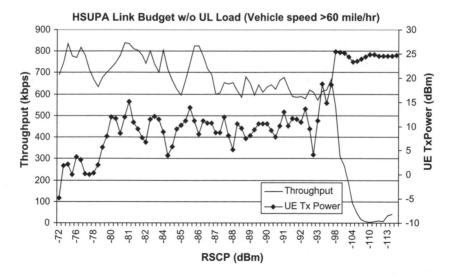

Figure 6.35   HSUPA link budget validation at high mobility (>60 miles/hr)

the cell range, therefore we believe that the results presented here represent a conservative case for HSUPA performance.

6.3.3.2 HSUPA Mobility Performance

Although soft-handover is supported in HSUPA, the absolute serving grant is controlled by the serving cell which is the same serving cell for HSDPA. Changing serving cells while in a HSDPA session causes low downlink data throughput in the soft-handover zone due to constant signaling and radio channel reconfiguration. Since the serving cell of HSUPA is controlled by the same procedure, this in turn will have a negative impact on HSUPA performance in areas with no dominant server. Figure 6.36 (a) and (b) show intra- and inter-Node-B HSUPA mobility performance. In the case of intra-Node-B where a dominant server is present in the SHO zone, the uplink throughput suffers some degradation but remains above 600 kbps. However, when the radio conditions were marginal and there was no dominant server in the SHO zone, as indicated in Figure 6.36(b), the throughput degradation was significant. A long service gap was observed during the test. The uplink bearer was downgraded to Rel.'99 several times during the cell transition.

The similarities between HSUPA and Rel.'99 channels lead to the belief that there should be no performance degradation in the SHO zone for HSUPA. However, as we can see in the results collected in the field, the HSUPA performance degraded even in areas where the radio condition was good (the RSCP was better than -80 dBm and only two pilots were in the active set).

There are several factors which differentiate HSUPA mobility from regular Rel.'99 soft-handover. In HSUPA, the E-AGCH, which provides the UE with an absolute value of the UL transmit power that can be used for data transmission, can only be established in the HSUPA serving cell (E-DCH serving cell) which is also responsible for the scheduling operation. This means that each time there is a serving E-DCH change, those associated control channels between the terminal and the serving cell will need to be reestablished before new data transfer can be scheduled for that user. In a SHO zone without dominant server, it can be expected that the serving E-DCH change (which typically follows the HS Serving Cell) will likely happen multiple times depending on how fast the mobile is moving. This in turn will cause the degradation of the HSUPA performance. Therefore, as was the case with HSDPA, a careful planning of the soft-handover areas, and of the hysteresis parameters for HS-DSCH cell change will have a significant effect on the performance of HSUPA mobility.

6.3.3.3 HSUPA Single Carrier Voice Plus Data

In this scenario we investigated the impact of HSUPA over voice, and vice versa. The tests were conducted in slow mobility conditions inside a single sector in cluster E. Up to 23 voice calls and up to three simultaneous HSUPA sessions were initiated without any degradation to the voice service. This result contrasts drastically to our previous observation in the lab environment, in which we found severe degradation of the voice users.

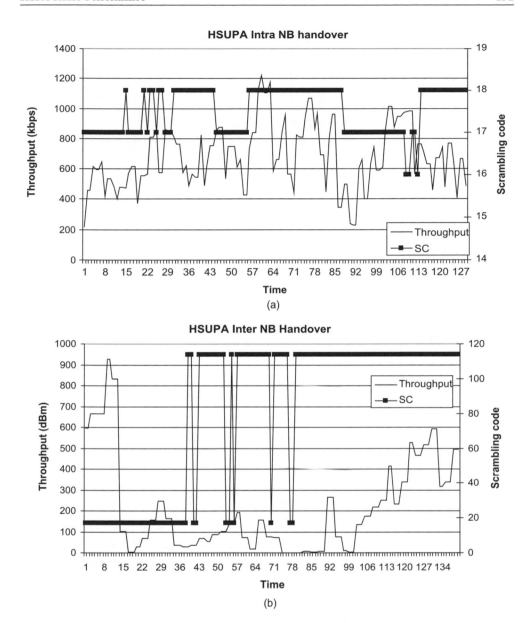

Figure 6.36 HSUPA throughput performance in SHO zone (a) Intra Node-B (b) Inter Node-B

Figure 6.37 shows that the overall sector throughput was 1 Mbps when no voice traffic was on the sector (limited by the baseband configuration of the Node-B). The three HSUPA sessions were in similar radio conditions and therefore achieved similar user throughputs. With 23 users the data capacity of the sector was reduced to 600 kbps.

The result was poor according to the expectations of the technology. The test results indicate that the HSUPA technology has not yet reached the degree of maturity achieved by HSDPA,

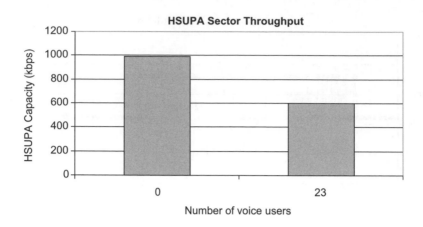

Figure 6.37 Effect of voice load on average UL throughput (3 HSUPA sessions)

with several opportunities for improvement in the short term including scheduling algorithms, reduction of UL power transmission and a more efficient resource sharing between voice and data.

6.4 Other Performance Considerations

In addition to the effect of the real environments in HSPA performance, there are some other factors which the operator needs to consider because we have found that they can have substantial impact on the data performance.

6.4.1 Terminal Device Performance

Although the 3GPP standard has defined most high level functions requiring terminal device's support, however, the actual algorithm or implementation varies from vendor to vendor. The discrepancy of the CQI reporting among different vendor's handsets is just one of the variations. The handset performance can be affected by design, integration and chipset selection etc. Figure 6.38 shows the HSDPA throughput of two Cat 6 devices under the same radio conditions. These two handsets are from two different handset manufacturers. So for operators, it is important to establish a benchmark to evaluate the different handsets' performance. This will help engineers and customer support representatives isolate problems when performance issues are raised by customers.

A difference in performance can ultimately be translated in a worse or less consistent end-user experience. On a different example, as the one shown in Figure 6.39, Device #1 experiences 40% shorter download times for large web pages. Different applications should be tested with each device, since just the throughput and latency alone may not provide all the necessary information about user perceived performance.

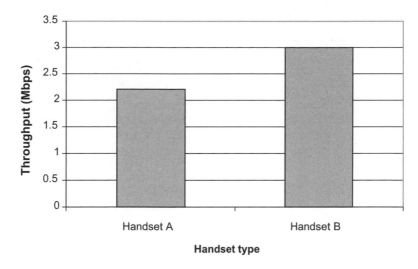

Figure 6.38   HSDPA performance of Category 6 handsets from different manufacturers under the same radio condition

6.4.2 Infrastructure Performance

On the network side, vendor specific implementations can almost be found in every node. Therefore operators should spend time to quantify the differences, especially those related to the performance. Figure 6.40 shows the latency measurements (lab) for two different RAN vendor platforms with the exact same topology and configuration.

Figure 6.40 is just one example showing that actual data performance can be related to a specific infrastructure or handset vendor's implementation; there are multiple of these

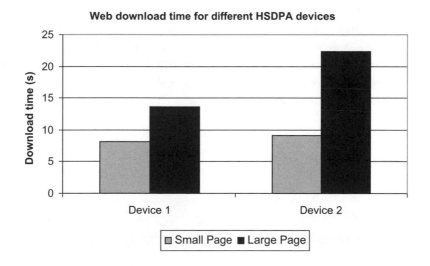

Figure 6.39   Web download times for two different HSDPA devices

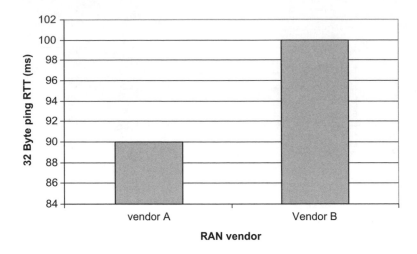

Figure 6.40   Latency performance for different RAN platform

examples throughout the previous sections when we analyzed the performance of HSDPA and HSUPA. In an operator's network, typically there are nodes from different vendors mixed together or in separate regions. It is important to benchmark the performance of each platform in the lab. The benchmarked performance information for each platform then should be passed to the field engineers to help the optimization efforts.

6.4.3 Application Performance

In order to remove the variations introduced by different applications when evaluating the radio performance of HSPA, FTP was used in the majority of the tests presented. From a RRM point of view, however, FTP is the simplest application for the RAN to handle since the data stream is steady and there is little channel switching involved during the transmission. In a real network, applications will be more diverse and challenging. Understanding the performance of each application to be supported by the HSPA network will help operators develop a proper marketing strategy for future data products. Figure 6.41 shows the uplink noise rise for a test with 10 simultaneous web browsing users in a sector.

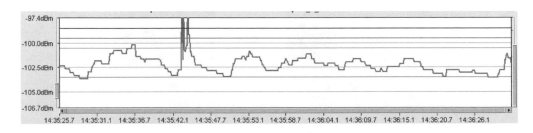

Figure 6.41   Uplink Noise rise with 10 web browsing users (default channel switching timer)

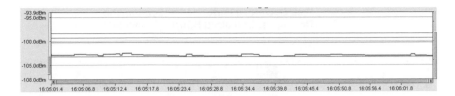

Figure 6.42   Uplink noise rise with 10 web browsing users (new channel switching timer)

In this test, each web user had specific traffic pattern: random reading time, random web page size and random starting time. The noise floor of the system was around −105 dBm. As can be seen, the uplink noise rise was as high as 8 dB. This was much higher than previous FTP tests which had bigger overall data payload.

During the tests, the system reached overload thresholds and users were downgraded or dropped when reaching 15 users. Those sudden interference peaks in the uplink were caused by the signaling of the channel switching back and forth from active to idle states during gaps in downloads. Since each web session had many gaps between page downloads, to save the system resource the RRM function will downgrade the user from Cell-DCH to Cell-FACH or Cell-PCH. When new web download is initiated, the user will switch back to Cell-DCH. These processes involve signaling between the UE and the RAN and causes excessive uplink noise rise. To fix this problem, the timer for the channel switching (Cell-DCH to Cell-FACH or Cell-PCH) had to be increased from 5 s to 30 s. Figure 6.42 shows an average reduction of 2 dB on the UL noise level after the parameter change.

However, this is achieved by sacrificing the system capacity since each user will occupy the resource longer. Figure 6.43 shows that the new parameters also achieve an improved HSDPA web browsing experience: up to 50% faster downloads with small web pages.

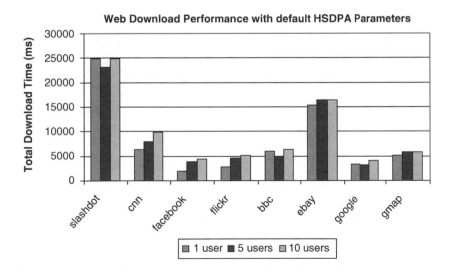

Figure 6.43   Web page download times for different pages and different amount of simultaneous users

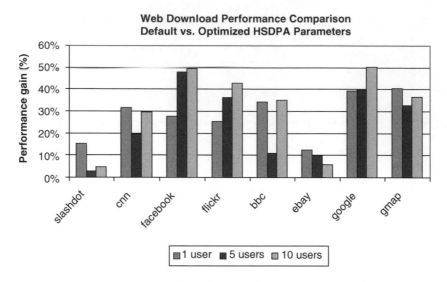

Figure 6.44   Web Performance improvement with new channel switching parameters

In Chapter 8, we will discuss the features which are introduced to provide continuous packet connectivity (CPC) while consuming less system resources. These features are very critical to the market success of the HSPA data offering.

6.5 Summary

The following are the main highlights from our lab and field test results:

- In general, the test results show the performance of HSDPA matching the expectations set by the technology for the given configuration. The results were conservative because of the practical constraints in the testing (e.g. limited Iub backhaul capacity, devices with up to five parallel codes, etc.).
- HSDPA delivered good, solid performance with stable throughputs up to the end of the coverage. At the edge of coverage, we recorded around 300 kbps throughput with 155 dB pathloss in different network environments.
- The vendor implementations of the RRM functionalities can make dramatic differences in terms of HSDPA and HSUPA capacity and throughput. The operator should identify the specific differences for the vendors present in their network.
- In the case of HSUPA the results are promising but in our opinion there is still room for improvement in the practical implementation of the associated radio resource management algorithms.
- Cell boundary areas (SHO zone) represent major challenges for both HSDPA and HSUPA. The operator should tackle this early in the network planning phase, and devote enough

optimization time after deployment to ensure dominant server coverage and smooth transitions between cells.

- With certain implementations the capacity of HSDPA and/or HSUPA can be significantly degraded with voice traffic and network interference. Our recommended approach to fully maximize both the voice and data capabilities is to deploy the services on different carrier frequencies if there are sufficient spectrum resources available.

References

[1] 3GPP Technical Specification 'Physical Layer Procedures (FDD)', 3GPP TS 25.214 v. 7.3.0 (Rel-7), p. 36.
[2] Tapia, P., Wellington, D., Liu, J., and Karimli, Y.,'Practical Considerations of HSDPA Performance', Vehicular Technology Conference, 2007. VTC-2007 Fall 2007 IEEE 66th Sept. 30, 2007–Oct. 3, 2007, pp. 111–115.

7

Capacity Growth Management

The traffic characteristics of 3G networks are different from those of older 2G networks because the improved data capability of 3G networks has enabled a new variety of mobile applications which are able to transfer higher volumes of data compared to 2G networks. With the wide acceptance of WCDMA data services around the world, capacity growth planning will be increasingly focused on data traffic, especially with the increased penetration rate of HSPA capable handsets in operators' networks.

One of the key questions that the operator will face when traffic volume increases is how to deal with congestion situations. In Chapter 5 we reviewed practical aspects of HSPA planning and highlighted that the best method to increase the capacity of a UMTS network that is not limited in coverage is to add new frequency carriers. However, the solution is not as easy as it sounds: there are several strategies that the operator may select when deciding to deploy a new carrier. Should the operator deploy the new carrier only in those sectors with capacity limitations? Is capacity the only criteria to deploy an additional carrier? When deploying the carrier, should it permit mixed voice and data or split the traffic? Section 7.1 will address these questions and provide a range of alternatives suitable for different operators' needs.

A second important consideration when dealing with traffic increase is how to dimension the transport network, in particular the last mile backhaul transport (Iub interface). As we indicated in Chapter 2, with the introduction of HSPA the main bottleneck is transferred from the air interface to the last-mile transport network. For this reason many operators deploying 3G networks have decided to change their transport media, from copper to fiber optic links, to avoid paying an excessive price for the transfer of increased data volumes. For those who still need to carefully plan their Iub resources, a tradeoff needs to be established between user performance and cost, which ultimately translates in Iub capacity. Section 7.2 provides guidance on how operators can factor in the different requirements to get an answer to their planning problems.

HSPA Performance and Evolution   Pablo Tapia, Jun Liu, Yasmin Karimli and Martin J. Feuerstein
© 2009 John Wiley & Sons Ltd.

7.1 UMTS/HSPA Carrier Deployment Strategy

In today's wireless environment, the multi-carrier deployment strategy is more and more driven by the growth of packet switched data traffic. The increased data capabilities of UMTS/ HSPA networks has enabled various data applications and increased traffic volumes. In order to satisfy the forecasted traffic demands, operators need to design a capacity planning strategy, which should consider the following factors: spectrum holdings, voice capacity growth, data bandwidth demand, data application profile and associated quality of service requirements. Operators should lay out a long term capacity growth plan and carrier deployment strategy (shared carrier or HSPA data-only) as early as possible to ensure that a solid initial network design is in place for future capacity growth, especially in networks where a data-only carrier will be likely deployed since it requires careful resource planning and each vendor's implementation can be different. In addition, the principles of designing a data-only carrier can be different from that of a traditional voice network, therefore, the focus on the design and optimization will be different for voice and data. Operators need to make sure that engineers understand the differences between voice and data services by providing proper training which in general should be planned well ahead of the 3G deployment.

As we reviewed in previous chapters, UMTS/HSPA networks allow voice and data services to be shared in a single carrier. An efficient assignment of the resources for both voice and data traffic is achieved through the radio resource management (RRM) function for the UMTS radio access network, as was discussed in Chapter 4. In particular, features like dynamic power allocation and dynamic code allocation greatly improve the efficient sharing of the resources.

The benefits of having voice and HSPA on a shared carrier are obvious: less spectrum usage, multi-RAB support (voice plus data connections simultaneously) and dynamic resource sharing between voice and data traffic. In the early stages of the 3G network deployment, when traffic loading is expected to be relatively low, the operator could start with a single carrier supporting both voice and HSPA. This allows the operator to launch 3G services rather quickly with a relatively low initial deployment cost. Since the network is lightly loaded, the data traffic carried by HSPA will have limited impact on the voice performance. With light traffic, operators have more flexibility to optimize the network and learn from the experiences, something that is especially critical for those operators who are introducing the 3G service and have limited operational practice with the UMTS/HSPA technology. Therefore from a short term perspective, starting with a single voice/data carrier is a recommendable solution for operators' initial deployment. Shared carrier deployment is analyzed in Section 7.1.2.

Although deploying a single carrier serving both voice and HSPA traffic provides much needed capital savings for operators during the early stage of the 3G deployment, when it is the time to add the second carrier, the decision on how to distribute voice and data traffic between these two carriers could vary for different networks. There is not a simple answer to this: the triggers to add the second carrier, and the configuration of this carrier can be different depending on the operators' data strategy and traffic distribution. Section 7.1.3 analyzes the benefits of deploying dedicated carriers for data traffic.

Long term capacity planning relies on many factors. The carrier deployment strategy varies from operator to operator or even from cluster to cluster. Therefore it is important to understand the available options and the benefits provided by each of those options. The purpose of this section is review the different factors affecting the decision to trigger a new carrier, as well as the various carrier deployment strategies analyzing the pros and cons of each option.

7.1.1 Factors Affecting the Carrier Planning Strategy

There are different factors which need to be considered when designing a capacity planning strategy. Table 7.1 lists these key aspects.

7.1.1.1 Traffic Growth

The forecasted volume of traffic is the primary consideration to be evaluated when designing a capacity planning strategy. If the traffic demand is low there will be no justification to increase the number of carriers. Also, the growth per traffic type is an important consideration: are voice and data traffic growing at the same pace? The service with the faster growth trajectory will typically determine the long term planning strategy.

Another important aspect to take into account is the fact that the traffic will not grow homogeneously throughout the network. In a real network, the traffic distribution is not uniform: there are always cells which capture more traffic and reach their capacity limits much sooner than others. The operator needs to realize that although some of those problems can be resolved by optimization techniques such as antenna down tilting, sector reshaping, pilot power adjustment etc, it is important to lay out the second carrier plan as early as possible because these solutions mentioned above sometimes could take months to implement due to site development related issues. Also, although RF optimization is recommended, especially when networks reach a mature stage, stretching the capacity of certain sectors purely based on RF planning can result in a less efficient network from an operational point of view. Well planned networks should grow in a holistic manner by adding new carriers to cope with capacity problems.

Table 7.1 Key factors for carrier planning

Key factor	Description
Traffic growth	Expected traffic volumes will be the main factor to determine the need for additional carriers. Uneven traffic distribution generates hot spots which may require adding carriers much sooner than planned.
Data marketing strategy	Type of application and expected data rate the operators want to offer.
Voice quality goal	Shared carriers can present degraded voice quality with higher loads. Lower codec rates may provide more capacity and maintain acceptable voice quality.
Radio environment	Unfavorable planning/radio conditions will result in lower cell capacity, requiring additional carriers sooner.
Mobility	Inter-frequency handovers needed to control transitions in cell boundaries.

7.1.1.2 Data Marketing Strategy

The operators' data marketing strategy plays a critical role in the carrier frequency planning process. The expected data traffic per cell can be determined by the data service plans offered by the operators to their customers. For instance, in general mobile devices have lower payload compared with data cards running on PCs or laptops. If the operators' business strategy is focused on delivering data services to smartphones, or broadband services to computer users, then the expected per cell data traffic will be much higher. In this case, it is possible that a second carrier will be needed due to the increased demand from the data services even when voice traffic is relatively low. On the other hand, if the operator's strategy focuses on handset-oriented applications, the expected data throughput per cell will be much lower. The triggering point for additional capacity may come at a later time. In this case, both voice and data could be the driving force for the carrier addition. This will depend on factors such as applications mix, device capability, 2G traffic migration and other factors which will vary from operator to operator.

7.1.1.3 Voice Quality Design Goal

The voice quality design criteria will also influence how the operator will cope with new capacity demands. For example, shared voice and data carriers may represent an impact on voice quality if data traffic is assigned too many resources (power). Another quality consideration is whether the operator wants to cover the increased voice demand with lower AMR codec modes, which can provide more capacity than AMR 12.2 kbps. There are many AMR codec sets defined in the 3GPP standards, the most common of which are the following:

- Narrow band AMR {12.2, 7.95, 5.9, 4.75 kbps};
- Wide band AMR {12.65, 8.85, 6.6 kbps}.

The number of voice calls a cell can support is a function of the hardware resources, Node-B power, codec rate, and network interference conditions. AMR lower codec rates have better coding protection and need less power to meet their quality target, therefore, the number of calls supported by a cell will be higher when a lower codec rate is used. Certainly, this is at the cost of sacrificing audio quality due to less information bits being transferred when lower codec rates are applied. For operators, the best practice would be to find the codec rate which can provide acceptable audio quality while consuming fewer power resources per radio link.

7.1.1.4 Radio Environment

In specific parts of the network, operators should be prepared to run short of capacity sooner than in other parts. To identify these areas, three different aspects have to be considered:

- *Terrain* has a direct impact on the network layout and neighbor cell interactions. Understanding the terrain features is the key to interference control. For a UMTS network, one cell's capacity is highly dependent on the interference generated by all its neighbor cells. In clusters with hilly terrain or high density tall buildings, interference control can be challenging. With higher overall network interference levels, the cell capacity will likely be lower in those cells.

- *Site density*. In dense urban areas, short site distances make interference control very challenging. Pilot pollution can be a constant impairment to the capacity.
- *Clutter profile*. Different clutter types (shopping mall, parking lot, office building etc.) all have different propagation characteristics. The extra penetration loss could translate into capacity loss.

7.1.1.5 Mobility

When deploying a new carrier it is important to consider the transition to the neighboring cells. If the adjacent cells have not been upgraded to the new carrier, then the mobiles will have to hand over to the common carrier before a cell transition can occur, which is achieved through the inter-frequency handover procedure.

There are some drawbacks associated with the handover between carriers: in the case of voice, this procedure results in degraded voice quality and increased probability of drops; in the case of HSPA the situation is even worse, because today most HSPA networks do not support compressed mode and therefore the packet call must first be downgraded to a Rel.'99 channel before attempting the inter-frequency handover.

The degraded inter-carrier mobility performance can be avoided if the new carrier is deployed in a cluster of sectors instead of just in a single hotspot. With a cluster deployment, mobility within the high-traffic areas (near the hotspot) will occur within the same frequency channel. Obviously, there will still be a boundary where the new carrier has not been deployed, but that boundary would have less traffic than the core of a hotspot and therefore the probability of failures would be reduced.

7.1.2 Voice and HSPA on One Carrier

In early UMTS/HSPA deployments, the 3G handset penetration rate most likely will be relatively low, and therefore the network traffic loading will be light for a certain period (between one or two years). By deploying a single carrier which supports both voice and data, the operators can put their main focus on achieving the UMTS/HSPA coverage on a large scale without committing more spectrum and equipment. This is especially critical for those countries like the US where operators can acquire spectrum before the incumbent operations are completely stopped. In such cases, the spectrum clearing process sometimes can be lengthy, and clearing multiple carriers can be unrealistic at the initial launching.

From a network deployment perspective, having a single carrier carrying both voice and data is relatively simple and straightforward to begin with. This situation will be optimal while the network traffic load is light; however, with the increased 3G traffic, optimizing the shared carrier will become more and more challenging, especially when voice and data follow different growth patterns in the network. More often, the operator will be facing the dilemma of meeting the design objective of one at the cost of the other. For a CDMA network, this is especially true considering the inevitably close relationship between the capacity and the coverage. Adding another dimension (data service quality requirements) to this problem will only make the situation more complicated. Several tradeoffs will need to be made among different services and their associated design objectives. These again, will depend on operators' market development strategy. For instance, if an operator's main goal for deploying UMTS is to provide high speed data services, then more system resources could be allocated for HSPA while allowing UMTS voice traffic to overflow to GSM when reaching target traffic loading criteria. Therefore, it is very important that clear design goals are set for all services supported by the shared carrier at the early stage of the network deployment.

7.1.2.1 Considerations on RRM Parameter Configuration

In order to effectively implement the shared carrier strategy, the operator should focus on the RRM functions. There are several factors which determine how the radio resources should be managed for a shared carrier. Depending on the strategy, the operators have the option of making different trade-offs between voice and data services. The following are the common factors which should be considered when configuring the different parameters governing the shared voice/data resources:

- expected data traffic growth rate;
- RF design criteria (interference management); and
- data application profiles (voice only, data only, voice+data).

Data Growth
Understanding the expected data traffic growth of the network is extremely important for operators to dimension system resources (power and base band) because in most systems, a significant amount of baseband resource has to be reserved to support HSPA. Likewise, the Iub backhaul bandwidth requirements will heavily depend on the HSPA traffic.

At an early stage of the UMTS deployment, operators may choose to have fewer resources dimensioned for HSPA for a certain period of time based on projected data traffic growth, which provides a relief on operational and capital costs immediately after the network launch. However, operators must have a long term plan for carrier expansion that incorporates flexible and quick methods to implement additional carrier frequencies once the network traffic grows.

RF Design Criteria
The configuration of the radio resources will in turn be conditioned by different factors, such as the desired voice and data quality, and the 3G handset penetration rate. If the operator experiences limited voice traffic due to low UE penetration during the initial deployment, then more resources could be dedicated to HSDPA traffic with no significant impact on the voice

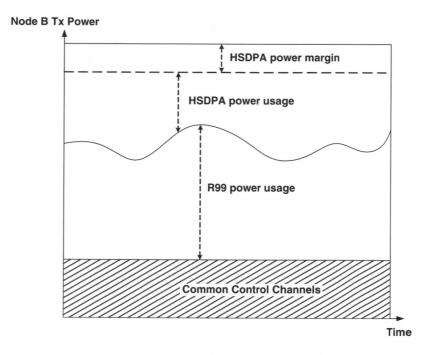

Figure 7.1 Power sharing between HSDPA and R99 traffic on a single carrier where Dynamic Power Allocation assigns the HSDPA power usage based on the Rel.'99 power usage

service quality. This could be accomplished by utilizing functions such as Dynamic Power Allocation (DPA) and Dynamic Code Allocation.

As it is shown in Figure 7.1, DPA is the typical scheme used by the Node-B to assign power resource for HSDPA. The goal of DPA is to allow more efficient power utilization when voice and data share the same carrier. Since HSDPA will be using the remaining power which is not used by Rel.'99 traffic, the cell will be fully loaded when there is HSDPA traffic regardless of the number of voice calls being carried by the cell. As a result, it can be expected that certain cells can be constantly at the full load condition even if the number of active users are low. The voice users will likely experience higher interference from their own serving cell once in a while depending on the data traffic volume. In early deployment stages, since the overall network loading is relatively low, the interference from neighboring cells may only contribute a small portion to the overall interference level seen by the UE; therefore, it is feasible to allow higher power usage for HSDPA in a cell. The impact of increased own cell interference introduced by HSDPA traffic will be limited as has been discussed in Chapter 6. Therefore, it is feasible to have a small HSDPA power margin, i.e. allowing the Node-B to transmit at near the maximum power level for HSDPA traffic without dramatically impacting voice quality.

Our test results collected in different trial networks have proven that in a lightly loaded network, the HSDPA power margin can be set to close to 0 dB without dramatically affecting the voice performance. In Chapter 6, we discussed the impact on BLER of voice calls when HSDPA and voice users were simultaneously served by the cell in an unloaded network. In one of the networks tested the HSDPA power margin was set to 0.2 dB at the time. The aggregate HSDPA cell throughput was consistently around 900 kbps with up to 22 voice calls being served at the same time, with 95% of the BLER samples below 4%. However, when the overall interference level of a network becomes high, HSPA performance should be sacrificed to maintain the quality of voice as the higher priority service.

Service Profiles

In addition to the capital savings and simplicity in the early deployment stages, having a single carrier supporting both voice and HSPA also has other benefits. Multi-RAB support is an important 3G feature introduced by 3GPP. By having voice and HSDPA on the same carrier, it is possible for the operator to enable richer services such as in-call video sharing, in-call location determination and other services, which in turn improve customers' experiences and create new revenue opportunities for operators.

Those services, however, may come with additional costs for network resources and need to be planned carefully. For example, voice and data may have different traffic patterns (e.g. busy hour, call duration, etc.) when independently offered, which allows the system to take advantage of the diversity of different traffic profiles, resulting in increased network efficiency. When these services are tied together, it will be more likely to experience resource contention among different applications, especially with services such as in-call video sharing, where the single user payload will likely be much higher. To support these multi-RAB services, the number of users supported by a cell typically will be much lower.

7.1.3 Data Centric Carrier

Allocating dedicated carriers for data traffic is not a new concept in cellular networks: that is, for instance, the principle behind the CDMA2000 1×EV-DO (Evolution-Data Only) technology, 3GPP2's 3G offering for data service, which has been deployed in many CDMA2000 networks around the world. In an EV-DO network, a dedicated (data-only) carrier needs to be assigned for data services, therefore simultaneous voice and data services can only be achieved through Voice over IP (VoIP) technology.

UMTS technology, on the other hand, is more versatile and can share the same carrier frequency between HSPA data with Rel.'99 voice. Therefore, the concept of 'data centric' carrier in HSPA has different connotations for other technologies, since that carrier could be serving voice traffic as well if the traffic demands require it. As it has been discussed in previous chapters, DPA allows HSDPA to use the remaining power which is not used by Rel.'99 voice. The power assigned to HSDPA can be adjusted on a per time slot basis and voice is always treated with the highest priority in terms of the Node-B power usage, so supporting voice and data simultaneously for one user is not a problem.

The main consideration is how to configure the carriers to support both voice and HSPA more efficiently. In the previous section, we discussed the pros and cons of having a single carrier supporting both voice and HSPA. It is obvious that this is not a universal solution which can be used for all scenarios facing the operators. Data centric carriers provide another viable solution which can fill in some of the gaps for operators' 3G growth planning challenges. In most vendors' implementations, HSPA can be enabled on per cell/carrier basis, which provides flexibility for operators willing to have separate voice and data services.

Compared to implementing a combined voice and data carrier, a HSPA centric carrier has the following advantages:

- *Accurate capacity growth planning.* Voice and data share different traffic characteristics: payloads, bit patterns, busy hours and service requirements all are different. Radio resource management is simpler when these two can be planned separately.
- *Higher, more stable data throughputs.* By having a carrier primarily dedicated to the HSDPA traffic (while still allowing a low level of voice traffic) a higher cell data throughput can be achieved with guaranteed power and code resources for HSDPA. Furthermore, the data performance of the cell will be more predictable since it is not dependent on voice traffic.
- *Simplified optimization scheme.* Because voice and data are on separate carriers, the operator can apply different design and optimization strategies for voice and data. This greatly decreases the complexity of the optimization work and reduces the operational costs.
- *Permits hot spot deployment of second carrier.* The HSDPA-centric carrier can be deployed as a hot spot solution in high density urban areas such as a shopping mall, campus area or office buildings where low mobility users with high bandwidth demands are expected. Under such a scenario, mobility normally is less of a concern for HSPA service, so continuous coverage for the new carrier may not be needed. However, if both voice and data are equally important on the new carrier, a cluster deployment (covering both the hot spot and the surrounding areas) of the new carrier will be needed to avoid constant inter-frequency handoff which degrades the voice performance and spectral efficiency.
- *Cost savings.* Typically, for each carrier supporting HSPA, the Node-B needs to reserve separate baseband resources. In a two carrier scenario, if both carriers support voice and HSPA, then dedicated base band resources are needed for each carrier to support HSPA. In a data centric scenario, only one set of base band resource needs to be reserved. This can provide significant capital savings for operators.

The disadvantages of having a data centric carrier are as follows:

- *Spectrum inefficiency.* A data centric carrier strategy requires more spectrum resources. The network will need to be equipped with two carriers at the initial deployment. This could be an issue in areas where operators have limited spectrum availability.
- *Mobility performance for HSPA.* In general, the coverage for the data centric carrier will be smaller compared with the voice carrier since not all sites will have the additional carrier. In some cases, the data centric carrier will only be deployed in certain hot spots to serve high density data traffic. When active HSPA users are moving out of the coverage footprint of the data centric carrier, the data call will have to be handed over to neighbor cells which may only have one shared carrier as shown in Figure 7.2. As has been presented, inter-frequency handover involves compressed mode which has been defined in the 3GPP standard but is not

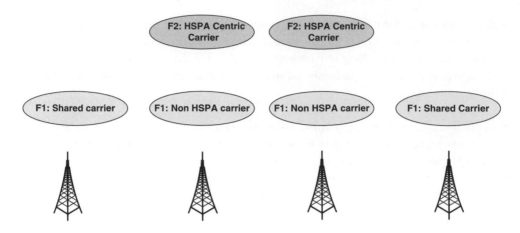

Figure 7.2   HSPA data centric carrier deployment in hot spot scenario

widely implemented for HSPA. Without HSPA compressed mode, the data call will be either: downgraded to Rel.'99 DCH first by triggering a radio bearer reconfiguration and then followed by the Rel.'99 inter-frequency handover procedure; or dropped and reconnected on the new carrier immediately. One way or the other, the data performance degradation will be noticeable to those moving users who are in active data sessions.

- *Increased cell reselections and cross carrier call setup assignments.* In idle mode, since there is no indication of the service (voice or data) that the user will be using next, the carrier in which the UE camps may not be the one in which the service requested by the UE is supported. Consider the network in Figure 7.2: in this case, a HSPA capable UE is camped on F1 and requests to establish a data connection, for which it is redirected to F2 during the call setup. And vice versa, a UE camping on F2 may request to establish a voice call, in which case it could be directed to F1. These considerations require operators to plan the camping strategy carefully to avoid excessive messaging created by constant cell updates. A simple approach is to only allow UEs to camp on one carrier, for instance on F1, for sites with a data centric carrier. In this case, only HSPA calls will be directed to the HSPA carrier. After the data call is completed, the UE will reselect to F1. The drawback of this solution is a possible degradation on call setup success rates because all call originations and terminations will be on one carrier. The interference level on this carrier will likely be higher, especially when traffic starts to increase, and as a result there could be increased call setup failures on F1 which could make the data carrier not accessible even though it has enough capacity. For the operators, it is very important to monitor the load on the carrier which is used for call setup control.

7.1.4 Factors Affecting the Shared vs. Data Centric Carrier Decision

Weighting these pros and cons of a shared versus a HSPA centric carrier solution, one should realize that there is not a universal solution. There are many factors contributing to the decision to deploy a HSPA centric carrier.

7.1.4.1 Spectrum Holding

First, the operator should evaluate the spectrum holding situation on a per market basis. Ideally, the HSPA centric carrier solution is more feasible for markets with at least than 15+15 MHz (downlink+uplink) of spectrum, i.e. three carriers. This allows separate voice and data network planning for a relatively longer period, since two carriers can be dedicated to either voice or data traffic depending on the growth rate. The HSPA centric carrier doesn't have to be deployed in every site in a cluster. Adding hot spots with data carriers is more flexible and does not impact the voice carriers.

Under certain conditions, an HSPA centric carrier could be implemented in places where only two carriers are available. For instance, in places such as an airport, coffee shop and public libraries, where the demand for data access may be higher than other places and voice traffic is relatively stable. However, this may turn out to be a short term solution if either voice or data traffic volumes cannot be supported by a single carrier. The operator will need to find solutions such as purchasing new spectrum, using lower codec rates for voice, or offload voice traffic to its legacy 2G networks (GSM).

7.1.4.2 Hardware Cost

The decision to add a data centric carrier should also consider the associated cost of the hardware. The new generation of Node-B's typically supports configurations of multiple carriers per cell without any additional hardware (power amplifier or baseband cards). The activation of the second carrier can be done remotely through the corresponding element manager. In these cases, since there is no cost associated to the operator, the decision to deploy a data centric carrier will be easier than for those operators who need to perform a site visit to modify the hardware configurations of the Node-Bs.

It should be noted, however, that there are often limitations on the maximum frequency separation between the carriers that can be supported by the power amplifier, which could require the deployment of new hardware even though the Node-B in theory could support higher capacity.

7.1.4.3 Traffic Projections

Traffic projections drive how the carriers should be allocated for data and voice. In markets where voice traffic still plays a dominant role and data services are not in high demand, the carrier addition should be mainly focused on voice, while supporting data at the same time. Under such conditions, sharing voice and HSPA on one carrier makes more sense. However, if voice started as dominant and the forecast projects significant future data traffic growth, the new carrier addition could be triggered by data. In this case the operators need to plan ahead and factor in the future data traffic volume in the early deployment.

Figure 7.3(a) and (b) show examples of different data traffic forecasts for a network. As was discussed in Chapter 6, in a mixed voice and data carrier, the data capacity will be limited to

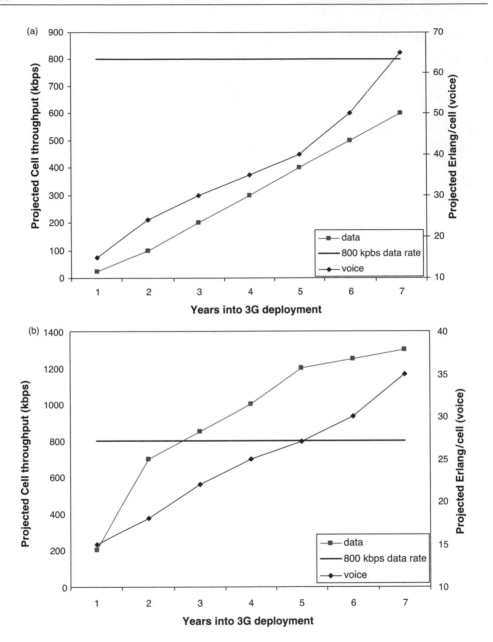

Figure 7.3 Example for different voice and data growth projections: (a) low data growth, and (b) high data growth

ensure the voice quality. In this example, we have assumed a UMTS cell capacity of 20 simultaneous voice calls plus 800 kbps of data traffic.

In case (a), the data traffic grows at a relatively slow pace and doesn't play an important role in triggering an additional carrier for the period under study. In this case, the operator can remain focused on voice capacity planning and share the extra capacity with data on all the carriers. The driving force for carrier addition will be voice.

In case (b), however, the projected data growth passes the threshold in the third year of the 3G deployment, and in the meantime the voice traffic will be growing at a slower pace. It is obvious that one carrier will be enough to support voice-only traffic for several years, but won't be enough if the carrier is shared by both voice and data even in the second year. In this case, deploying a HSPA centric second carrier is a better solution since the main capacity growth will come from the data traffic. Realizing that capacity demand for data will continue to grow and outpace that of the voice, applying data centric carrier strategy serves better from a long term plan point of view under such a scenario.

7.2 Data Traffic Profiling and Network Dimensioning

The growing importance of data traffic versus voice has been reshaping many operators' strategies on network planning. Backhaul dimensioning is certainly no exception. There are many fundamental differences between voice and data traffic as has been discussed in earlier chapters of this book. These differences shown in Figure 7.4 lead to different approaches for network design, optimization, capacity planning and resource dimensioning. The main objective of this section is to model the real data traffic profiles, and their impacts to the resource dimensioning of operators' 3G networks.

7.2.1 Traffic Profiling

In traditional cellular networks the main traffic sources driving the dimensioning exercise have been voice services and simple circuit-switched data applications such as SMS. This trend has

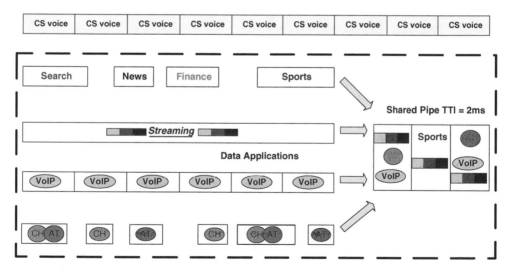

Figure 7.4   Different traffic characteristics between voice and data

been slowly changing with the introduction of new network technologies and feature-rich handsets that offer a multitude of mobile applications.

The normal approach taken by operators when dimensioning their networks has been to focus on voice traffic, and treat data as 'best effort', or in other cases as circuit-switched data with a constant bitrate requirement. While the dimensioning of voice traffic is typically based on Erlang tables, which provide a useful mapping between the number of channels that need to be dedicated to the service and the expected Grade of Service (GoS) – typically blocking – there is not an analogous method to dimension the network for data traffic. The reason is that 'data' is not a service per se, but a multitude of different services that need to be analyzed separately but treated as a whole entity.

The traditional approach for data dimensioning has focused on the capabilities of the air interface, neglecting many important aspects such as traffic quality of service (QoS) requirements, application associated traffic patterns, efficiency of the data scheduler, etc. Such an approach is referred to in this section as the 'full buffer' model (which assumes that the air link continuously transmits full rate streams to active users), and does not represent a good approximation for networks such as HSPA which can support much higher data rates on the radio interface.

The following is an example of how application profiles can drive the network dimensioning. Assuming there are four types of applications, as depicted in Figure 7.5, running simultaneously by a group of users, the backhaul requirements can be different depending on how the QoS is handled on the network side. Case (a) is what we call Peak Throughput Dimensioning. Under this approach all applications are given the same priority and each of them needs to be delivered when packets arrive. This kind of approach is very typical among operators when they are dealing with network dimensioning for data applications at an early stage.

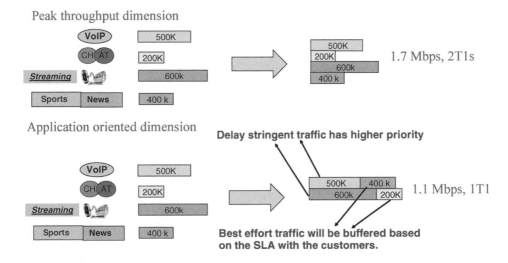

Figure 7.5 Backhaul dimensioning for different application profiles (a) Peak Throughput Dimensioning method and (b) Application QoS considered

This approach results in over-dimensioning of the network elements and interfaces, which increases the equipment and operational costs.

It can be clearly seen in case (b), when different packet delay requirements for applications are applied in the planning, then the required backhaul resource can be reduced significantly as compared to the Peak Throughput Dimensioning method. In a HSDPA system, data traffic from different users is delivered by the shared channel on a per TTI basis, which is controlled by the HSDPA scheduler. In early releases, HSDPA schedulers are simplistic and have minimum or no QoS awareness, so all data traffic is treated equally. This certainly will introduce inefficiencies in the utilization of the system resources and degrade user experiences for certain type of applications, however this would only be a temporary situation. With the introduction of more HSDPA traffic classes (conversational and streaming) and more sophisticated schedulers with QoS awareness, the system will be able to capitalize on the benefit provided by the diversity of the data traffic from different applications. For a scheduler with QoS awareness, packets from applications with stringent delay requirements, such as streaming and VoIP, will always have higher priority and will be delivered first whenever the resource is available. In the meantime, those background data traffic such as email and music download which are not time critical can stay in the buffer for a longer time and be delivered in TTIs later without causing congestion of the system and poor user experience.

One of the key elements to perform an efficient dimensioning of the network is to gather relevant information about the data traffic volumes and characteristics of the specific applications that are served by the network. In many occasions the only available data traffic information is the estimated average monthly data usage per user, which makes it very difficult for operators to dimension the network accurately. When the traffic characteristics and priorities are not considered in the capacity planning process, there will be a high likelihood of over-dimensioning the network resources. Therefore it is important to define methods to obtain new KPIs from the network that provide information about the types of applications being served and their traffic characteristics.

Figure 7.6 depicts a diagram with the main steps to perform a data dimensioning exercise based on QoS requirements for different applications. Note that another important input to the process is the definition of what 'acceptable' user quality mean to the operator. This is a very subjective term and different operators may have different views. The expectations also depend on the type of network (wired vs. wireless, etc.). Table 7.2 provides an example of these 'acceptable' user quality criteria.

Three main contributors for a successful capacity growth planning are the following: (1) traffic profiling, (2) availability of a QoS-aware scheduler and (3) network optimization. With more and more data applications being developed for mobile devices, it is important for operators to understand the nature of these applications and create the appropriate planning strategy to meet the challenge of the increasing data demand. Scheduler performance and network optimization have been covered in other chapters of this book. Here we will focus on analyzing the data traffic profiles and the impact to the backhaul dimensioning. A traffic model is used to assist this analysis.

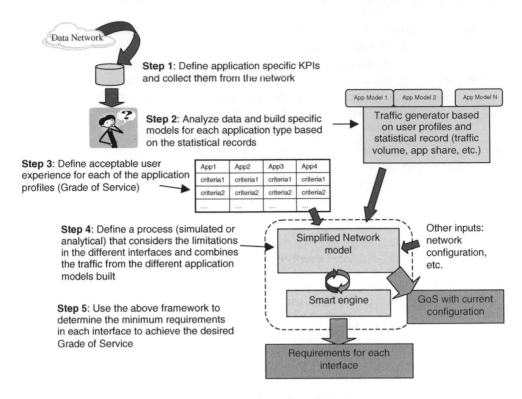

Figure 7.6   Diagram of the dimensioning process

7.2.2 Data Traffic Models

The increased penetration of smartphones in cellular networks has provided a golden opportunity for the software development community. People sense the potential of what a high speed network with mobility support and great coverage can deliver and have been trying to capitalize on this by developing all sorts of mobile applications. Since many of those applications are client-type software that can be downloaded by the end users, the operators have limited control over the content being delivered. This presents a great challenge to

Table 7.2   Examples of quality criteria defined per application

Application	End user criteria
Web browsing	Small page (<50 kB) in less than 5 s
	Medium size page (<100 kB) in less than 10 s
	Large size page (<500 kB) in less than 30 s
	Others: no expectation
MMS/FTP upload	Send pictures (up to 3 MB) in less than 30 s
Streaming	< 1 rebuffering event every 60 s
. . .	

Table 7.3   Key parameters for traffic model generator

T	The simulation time (traffic observation time)
K	The total number of active users during T
App_prob	Application occurrence probability vector (proportion of time of each application)

operators' network planning and makes it very important for operators to understand the nature of the data applications and their impact to the capacity planning.

This section presents examples of traffic models for the most typical applications found in mobile networks today, including web browsing, video/audio streaming and email/file download. In addition, since typically the data traffic is asymmetrical and biased toward the downlink, all modeling assumptions are made based on the downlink traffic pattern.

As indicated in Figure 7.6, the data traffic models should be embedded in a traffic generator that will create different users according to the application profiles indicated by the operator. Each application will have its own packet distribution characteristics and delay requirements. These can be configured through the parameters defined in the model. Table 7.3 shows the key parameters for the traffic generator. In reality, K and App_prob will vary in a real network, however with a short enough T these can be considered as constant.

For each user, the traffic generator will randomly pick an application session and call the specific application function/generator. The traffic for an individual application as well as the aggregated traffic can be created for each time interval. Then it is up to the scheduler to determine how each of those packets which belong to different applications should be delivered. As we mentioned, each application may have different packet distributions and arrival patterns, so it is important to understand the details of each of the traffic profiles.

7.2.2.1 Web Browsing Model

A web session is composed of multiple page downloads separated by a random 'reading time' in between the pages, as indicated on Figure 7.7. Each of these page downloads can be treated as an individual packet call.

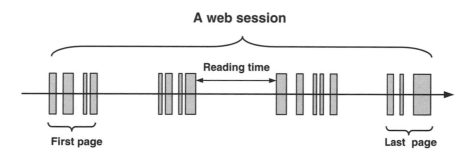

Figure 7.7   Web browsing packet arrival pattern

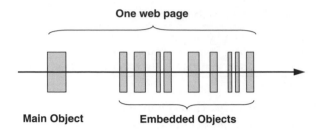

Figure 7.8   Traffic pattern within one web page

Within each web page, there are multiple objects. The very first one is the main object and each of the objects referenced from the main objects is an 'embedded object', as shown in Figure 7.8. Images are typical embedded objects in the main object HTML script.

The HTTP traffic generator basically is a random event generator using the parameters defined by Table 7.4.

7.2.2.2 FTP Model

In FTP applications, a session consists of a sequence of file transfers, separated by reading times (see Figure 7.9). The two main parameters of an FTP session are:

- File size: the size of a file to be transferred. The file size distribution typically is a truncated lognormal distribution.
- Reading time: the time interval between end of download of the previous file and the user request for the next file. Reading time typically follows exponential distribution.

Similar to the web traffic generator, the FTP traffic generator applies the probability density functions for the file size and reading time and generates FTP traffic for the time period defined by the user. Compared to the web traffic, the packet size distribution is more uniform.

7.2.2.3 Streaming Video Model

Streaming video traffic has stringent delay requirements. Video packets arrive at a regular interval T, and the number of frames per second (FPS) determines the video quality. During the

Table 7.4   Parameters for HTTP traffic generator

Variables	Distribution
Size for the main objects	Truncated Lognormal
Size for embedded objects	Truncated Lognormal
Average number of objects	Truncated Lognormal
Reading time	Exponential
Parsing time	Exponential

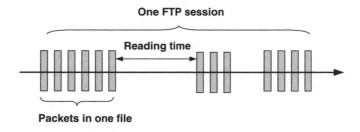

Figure 7.9   Traffic pattern for FTP applications

transmission, each video frame is decomposed into a fixed number of slices and each is transmitted as a single packet. On the receiver there is a buffer that stores packets before they are displayed on the screen. The buffer size is a function of the FPS and available bandwidth. The bigger the buffer size is, the longer the initial waiting time for the video. Figure 7.10 shows the packet pattern in a streaming session.

The traffic generator has the following control parameters:

- session duration;
- frames per second;
- slice size: typically a truncated Pareto distribution; and
- encoding time: packet/slice inter-arrival time, typically a truncated Pareto distribution.

The video streaming traffic generator uses the packet distribution directly and applies the coding time to each packet/slice. The buffer window, as mentioned, is a configurable parameter in the model.

7.2.2.4 WAP (Wireless Application Protocol)

For mobile devices with limited processing capability, WAP is typically used for web browsing applications. Since the objects are modified to fit the device capability, the object size

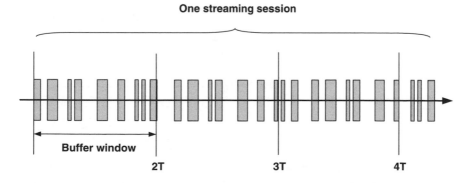

Figure 7.10   Traffic pattern for streaming applications

Table 7.5 Parameters for WAP traffic model

	Distribution
Object size	Truncated Pareto
# of objects per request	Geometric
Interval between objects	Exponential
WAP gateway response time	Exponential
Reading time	Exponential

Table 7.6 Configuration of the HTTP traffic model in the example

Variables	Average value
Size for the main objects	10 KB
Size for embedded objects	5 KB
Average number of objects	30
Reading time	20
Parsing time	3

distribution is different from that of the HTTP traffic model. Table 7.5 lists the parameters for the model.

7.2.3 Data Traffic Modeling Case Study

This section presents an example study for a HSDPA network with web traffic only. The objective of the exercise is to determine how many simultaneous users can be served with a single T1 (1.54 Mbps nominal bitrate). The model considers that several web sessions will be generated in the sector according to the parameters indicated in Table 7.6.

The results from the simulation are show in Figure 7.11. According to the results, the network is able to serve up to 20 users with a single T1 with very little degradation for small web pages, and around 20% degraded performance with large pages. Ultimately, as mentioned in earlier sections, the operator needs to define an acceptance criterion for web browsing

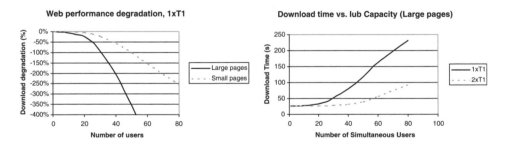

Figure 7.11 Results of several dimensioning simulations. Left: performance degradation with increased number of users (1 × T1); Right: web download times for a large web page (300 KB) for different backhaul assumptions (1 × T1 and 2 × T1)

quality. If the criterion is set to a target download time shorter than 25 s, the results indicate that a single T1 is enough for up to 20 simultaneous users, and two T1s would be needed to support up to 40 simultaneous users (see Figure 7.11, right).

It should be noted that the numbers used here are just an example to illustrate the concept. However, the important message to be delivered is that resource dimensioning for data traffic is a multi-dimensional problem. Operators need to look at all aspects of the network: application profiles, network loading and QoS requirement when dealing with the resource planning for data traffic. The cost between application oriented dimensioning and the peak throughput planning can be substantially different.

7.3 Summary

Capacity growth management is one of the central elements of a wireless carrier's business strategy. With the introduction of high speed data to the network, data applications are becoming the driver for operator's revenue. The steady decrease of average revenue per user (ARPU) in the past several years has led operators to search for new revenue streams. Data applications, with the richness in content delivery, are no doubt the next evolution of wireless services. The improved air interface capabilities have shifted the focus from improving the air link data capacity to managing the resource dimensioning efficiently. It is a complicated task which requires operators to establish a new capacity planning strategy which is application oriented and tied to the QoS of the applications.

The following are the main takeaways from this chapter:

- Operators should define a carrier planning strategy well in advance of the need to add new carrier frequencies to minimize hardware modifications and site visits.
- Shared voice and data carriers are a simple solution but not always the most efficient one.
- The forecast of data growth is a key element in planning the addition of new carriers. A significant amount of expected data traffic would justify a strategy based on data centric carriers.
- The capacity planning strategy should also define how the additional carriers will be installed, in hot spots or clusters.
- Another important aspect of capacity growth is the network dimensioning in particular the Iub backhaul and baseband resources. Data traffic simulation models can be useful in the dimensioning exercise.
- In the case of increased data volumes, the dimensioning exercise should be based on application Grade of Service rather than 'best effort' or 'peak dimensioning' to optimize the efficiency of the system and achieve the best user performance.

References

[1] Naguib, A. et al., 802.20 Evaluation Methodology' C802.20-03/57, IEEE, 2003.

8

HSPA Evolution (HSPA+)

With converged devices and IMS-based applications, the communication industry transitions towards data-only networks, also known as 'All-IP' networks. In these networks both voice and data services can be offered with similar or better quality than what is experienced on today's cellular networks, with significant cost advantages derived from the utilization of Packet Switched (PS) backhaul, together with a simplified architecture and network management. Because of the pervasive adoption of the internet, All-IP networks are cheaper to build, grow and operate than typical mobile networks, and offer additional benefits such as reduced time to market for new services and functionality, allowing for increased responsiveness to customers.

The migration trend towards an All-IP network is already prevalent in landline networks and is now accelerating in wireless, especially with the introduction of WiMAX and LTE, air interfaces which no longer support circuit-switched operation. As core and radio networks transition to IP, it becomes possible to move from circuit switched voice to packet switched voice.

Voice over IP (VoIP) is one important piece of the puzzle in the All-IP transition, because it will permit the unification of the core network into a single, packet data core. Furthermore, offering the voice service over IP in a HSPA network has additional benefits such as the possibility of introducing new, richer converged voice, data and multimedia applications.

In this chapter we discuss the enablers that 3GPP provides via HSPA+ in order to support new data services. We provide an overview of the changes introduced in the Rel.'7 and Rel.'8 standards, first with a review of the new radio features in Section 8.2 and then of the associated architectural changes in Section 8.3. At the end of the chapter we summarize the benefits of HSPA+ by using a practical example of voice and data convergence with the deployment of Voice over IP. We review requirements that an operator needs to consider in order to deploy converged services where voice and data are no longer separate applications; rather voice is just another data application.

HSPA Performance and Evolution   Pablo Tapia, Jun Liu, Yasmin Karimli and Martin J. Feuerstein
© 2009 John Wiley & Sons Ltd.

8.1 Standards Evolution

The standardization and early commercial availability of IEEE's WiMAX technology in 2004 and the Mobile WiMAX version in 2005 triggered a heated debate inside the main cellular standards bodies that saw how an external player (IEEE) could potentially be positioning its technology as the next evolution step for wireless communications. WiMAX offers a spectrally efficient radio access technology based on Orthogonal Frequency Division Multiple Access (OFDMA). WiMAX also introduced spatial multiplexing (MIMO) to cellular applications, and because of the reduced frame structures and flatter architecture the technology can achieve reduced network latency. The WiMAX network architecture is simplified compared to typical cellular network architectures, with only two layers – access and gateway – which could represent significant CapEx and OpEx savings for the operators.

In 2005 3GPP initiated standardization of an alternative evolutionary technology, known as UTRAN Long Term Evolution (LTE) or Evolved Packet System (EPS). The technology is very similar in concept to WiMAX and represents a signficant evolutionary stem from the GSM/ UMTS family tree. The goal of operators to deploy such technologies would be to achieve significantly higher peak data rates, spectral efficiency and lower latency, all with a more simple network architecture and much lower infrastructure and deployment cost than the existing networks. All of the proposed architectures are packet data only networks (All-IP), and therefore any voice communication would have to be done through the Voice over IP (VoIP) service. Chapter 9 provides insights into LTE.

LTE standardization has attracted much R&D effort within the industry, which in turn, triggered a reaction from some UMTS operators who felt that UMTS/HSPA still had room for improvement. These operators were not convinced that they should jump to LTE in the short term. A plan for HSPA evolution was proposed by 3G Americas in 2006 through an initiative that would later be called HSPA+ [1]. At the time of the submission, 3GPP Rel.'7 standardization work had already started, but the content was very light and no significant radio link performance improvements were envisioned. 3G Americas contended that Rel.'7 needed to be richer in content to improve capacity and performance similar to enhancements within Rel.'5 and Rel.'6, which had brought to life HSDPA and HSUPA respectively. HSPA+ proposals included a set of improvements that would help the HSPA (HSDPA+HSUPA) technology perform closer to what was being promised by LTE, thus providing a longer life to UMTS. This proposal would have a similar effect on UMTS as EDGE did for GPRS, which significantly extended the 2G lifetime.

Rel.'7 standards work started in 2005 and was focussed on features that enable converged services such as VoIP, push to talk and video sharing which have been made available with the deployment of the IP Multimedia System (IMS) by operators. Between 2006 and 2007, significant progress had been made in the UMTS standard development, and the new work items were closed in March 2007. Some of the HSPA+ features were already introduced in Rel.'7, while others are currently standardized under Rel.'8. HSPA+ includes a series of enhancement to the HSPA radio interface such as higher order modulation, dual carrier support and interference-aware receivers. It also introduces an evolved RAN architecture which approaches the flat architecture concept envisioned by LTE and WiMAX.

8.1.1 Radio Evolution

Rel.'7 enhancements cover new features in core network architecture including IMS as well as in the radio access network. On the radio access interface, the introduction of MIMO and higher order modulation further improve the peak and average throughput of HSPA. The new interference-aware receivers not only are able to estimate the own-cell interference, but also can detect the other-cell interference and attempt to cancel the interference contributions. This improves the spectral efficiency as well as the data performance at the cell edge.

The performance benefits from Rel.'7 are multifold. In addition to the capacity improvements introduced by new radio features such as uplink Continuous Packet Connectivity (CPC) and DL MIMO, the latency reduction will be substantial compared to the Rel.'6 platform [2]:

- Round Trip Time improvement from <100 ms to <50 ms;
- Improved Packet Call Setup Time reduction from ~1000 ms to <500 ms; and
- Improved Control Plane Latency, Dormant to Active, reduced from ~1000 ms to <100 ms.

In existing Rel.'6 HSPA networks, latency has been improved with the introduction of HSUPA. Under ideal conditions, HSUPA latency can be in the order of 60–70 ms. However, in real network deployments with routing delays and wide distribution of application servers, the latency will be much higher, which may prevent offering delay stringent applications such as VoIP. With the development of new applications (e.g. online gaming, online chat rooms, etc.) on mobile devices, more and more traffic will fall under the profile which requires low latency and low bandwidth. Therefore the improvement in network latency will be as critical as the increase in the peak data rate, if not more important.

An increased number of simultaneous active data users is also an enhancement in Rel.'7. It is an important feature for the system to support 'always on' applications with low bandwidth demands.

Rel.'8 standardization work is still in progress, but it will introduce new features such as Dual Carrier for data services, Circuit Switched over HSPA and Fast Cell Selection for HSPA.

8.1.2 Architecture Evolution

For core network and IMS features, Rel.'7 makes IMS the common platform for both wireless and wireline services. It allows easy integration of application platforms that enable network operators to create and deliver simple, seamless and secure multimedia services to their customers. VoIP and voice call continuity (VCC) allow operators to have the option of migrating voice services from the circuit switched (CS) domain to the packet switched (PS) domain, while maintaining seamless mobility between the IMS-based VoIP services and the conventional CS services.

Other significant additions are the introduction of a flat GPRS architecture concept (GGSN direct tunnel), which further reduces network latency and improves network scalability of packet data traffic, and the design of a new Policy and Charging Control (PCC) architecture which allows operators to perform advanced dynamic QoS control and charging for packet data services.

8.1.3 Vendor Ecosystem

With progress in the 3GPP standards, infrastructure and handset vendors have been pushing for the implementation of new Rel.'7 features in their platforms. For instance, a significant effort has already been put into the improvement of the HSDPA receivers by both the chipset and equipment vendors, even before the Rel.'7 standard was finalized. Some vendors have been very active in implementing the flat core architecture for HSPA+, while others are more focused on the radio interface improvements. Since some of these implementation and studies are done in parallel to the standards development, it is very beneficial to the standard development by contributing findings through real-world implementations.

For operators, choosing the HSPA+ evolution path does not mean that they will not eventually hop onto the LTE train. Since many vendors have been working on both the HSPA+ improvements and LTE/EPS development at the same time, the migration from UMTS/ HSPA can be simply a software upgrade for some vendors' products. For instance, many vendors' latest Node-B designs have already factored in the LTE standard requirements. Migration from the current technology to the next generation is becoming less painful for operators. It gives operators the option to leverage the existing HSPA ecosystem, while waiting for the ecosystem for LTE/SAE to become mature.

8.2 HSPA+ Radio Enhancements

8.2.1 MIMO

MIMO stands for multiple input, multiple output, which in general refers to a wireless system with multiple antennas at both the transmit and receive ends of the link. There are four different types of transmit-receive antenna combinations, as depicted in Figure 8.1. In most operators' networks, single in and multiple out (SIMO) is the configuration being widely used on the uplink where two receive antennas are used for Rx diversity purposes. On the downlink side, due to the form factor and battery consumption limitations, receiver diversity is only implemented in data cards. For many handsets, there is only a one single receive antenna, therefore in most cases the downlink is a single input, single output (SISO) system. Recently, there have been increased efforts in reducing the handset battery consumption and bringing in new antenna designs to the handsets, and it is expected that handsets with Rx diversity will be launched in the market in the near future.

Depending on the implementation, MIMO can be divided into several different categories. Transmit diversity uses multiple Tx antennas to counter the effects of channel fading. It

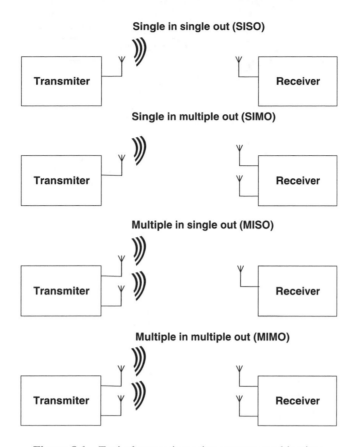

Figure 8.1 Typical transmit-receive antenna combinations

improves the cell average throughput by applying space-time coding schemes to a single data stream. However, the peak data rate for the individual user is not increased since the information bits from those multiple antennas are from the same data stream. Spatial multiplexing (SM), on the other hand, can increase the peak throughput by transmitting multiple data streams through different antennas. In 3GPP Rel.'7, the focus of MIMO is mainly on SM.

There were many MIMO proposals in the early standards work, however only two proposals were finalized in Rel.'7 considering realistic deployment scenarios which meet the operators' expectations: Per-Antenna Rate Control (PARC) and Double Transmit Adaptive Array (DTxAA). In 2006, it was agreed to standardize the MIMO as following in Rel.'7:

- PARC is selected for TDD system;
- DTxAA is selected for FDD system. The weights are signaled on the HS-SCCH in the downlink.

Figure 8.2 Shows the diagram of DT×AA MIMO for FDD [3].

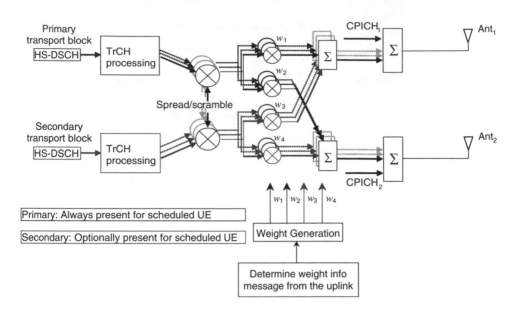

Figure 8.2 MIMO downlink transmitter structure for HS-PDSCH (UTRA FDD) [3] © 2008 3GPP

The performance gain provided by MIMO is highly dependent on the network geometry. In general, spatial multiplexing is most effective in high geometry areas where inter-cell interference is low. In a typical wireless network, the areas which meet the criteria to transmit with spatial multiplexing are limited. The capacity gain of MIMO can be better realized in microcell type of radio environments such as malls, airports and convention centers. These areas also provide the rich scattering environment that MIMO uses to improve peak data rates through the parallel MIMO channels. In areas where the geometry is low, using single data streams with space time coding (STC) will provide better performance. Industry studies show between 17% to 28% capacity gain in microcell environments, and around 15% in macrocells.

Although MIMO has been put into Rel.'7 as one of the main radio interface improvements, it hasn't been widely accepted in the industry as a must-have feature for HSPA+ due to the limited capacity gains. In the meantime, features such as advanced receivers for HSPA have been proven to be far more beneficial in improving the data performance while introducing limited complexities to the network. In addition, the concept of having multiple transmit antennas (and multiple power amplifiers too) generally doesn't sell well in the operator community. On the downlink side, it means that more space may be needed on the tower or facilities where antennas are mounted (one exception would be 2x2 MIMO with cross-polarization antennas). This has cost implications on both the operation/maintenance and site development sides. In addition, the added complexity in the structure (cell tower, rooftop etc) could also cause lengthy delays related to the local zoning process which if possible, operators always try to avoid. On the uplink, adding one more transmit RF chain means more power consumption and shorter battery life for the handset.

Based on these considerations it is expected that MIMO will be typically deployed in specific sectors with high bandwidth demands, typically hotspot in dense urban areas, rather than a network-wide deployment.

8.2.2 Higher Order Modulation (HOM)

In Rel.'6 HSPA, the HSDPA link adaptation uses two modulation schemes combined with different coding rates on the downlink. On the uplink, only QPSK is used on HSUPA. One drawback of using higher order modulations techniques such as 64QAM is the fact that these require extremely clean and strong signals that cannot be achieved across all of the cell coverage area. Figure 8.3 illustrates the utilization of 16QAM in two different drive test routes, the first one in an urban Rel.'6 UMTS cluster with 0.4 mi of inter-site distance (left) and the second in a suburban environment with 1.2 mi of inter-site distance (right).

In the case of the urban environment the weighted average percentage that the UE is in 16QAM mode was around 60%, and around 50% in the case of suburban environment. In the case of higher modulation such as 64QAM, the potential area of effectiveness will be further reduced and the gains are not expected to be dramatic. Note that these values were collected without traffic loading in the network and with all the cell power allocated to HSDPA, so it can be expected that this ratio will be reduced when the network traffic load goes up.

The feasibility of using 64QAM was heavily investigated by 3GPP as early as 2001 and wasn't adopted in earlier releases for the reasons mentioned above. However, recent studies by 3GPP Working Group (WG) 4 on new advanced receivers have renewed the interest in 64QAM in the wireless community. Figure 8.4 shows the link level simulation for 64QAM done by 3GPP WG1 in 2006 [4]. It indicates that for the advantageous users (90th percentile), 64QAM can increase the bit rate up to 35% (depending on the load) for a Pedestrian A channel profile, but only slightly (around 10%) for a Typical Urban (TU) profile. The capacity gains are not considerable: around 5% for fully loaded cells, and between 10% and 20% for partially loaded cells.

With the introduction of advanced receivers in Rel.'7, the possibility that the system runs into peak rate limitation will substantially increase, especially in 'contained' RF environments

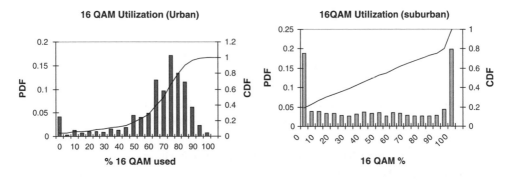

Figure 8.3   Percentage of 16QAM usage in an urban cluster (left) and suburban (right)

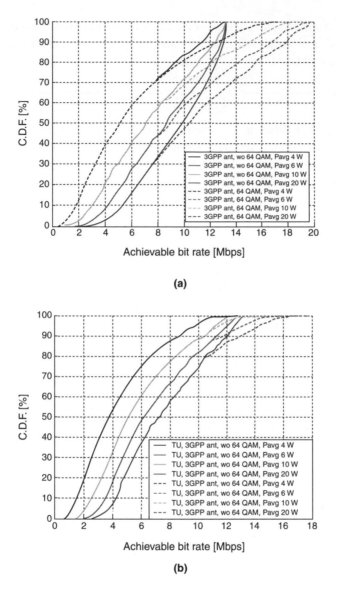

Figure 8.4 64QAM link level simulation for different network loads in (a) Pedestrian A, and (b) Typical Urban radio channel models © 2008 3GPP

where network geometry is high. Therefore it is a logical decision for the standards group to include 64QAM as part of the Rel.'7 features to leverage the performance benefits provided by the new receivers.

Working together with MIMO, 64QAM can provide added flexibility for operators to increase the HSDPA capacity; however this possibility was not permitted in the standards until Rel.'8. As discussed in Section 8.2.1, deploying MIMO generally requires site modifications since multiple transmit antennas are needed. In many cases, the operators don't have the option of doing so because of issues related to zoning, cost, structural and environmental concerns. Having 64QAM as one of the options helps operators to add capacity quickly in areas with high density traffic such as malls, airports etc., without site development work involved. In many cases 64QAM can be implemented as a software upgrade, without any hardware impacts to the network.

In addition to 64QAM on the downlink, 16QAM is added for HSUPA. The uplink peak rate is increased from 5.76 Mbps to 11.5 Mbps. Again, this feature will benefit from the new generation of Node-B receivers, which incorporate Interference Cancellation or HSUPA Equalizers. On the other hand, the deployment of 16QAM in UL could lead to increased UL interference. During the early deployment of HSUPA with Rel.'6 terminals (QPSK only on the E-DCH) it has been observed that the mobile transmit power is at the high end, even in locations where radio conditions are good. Adding higher modulation on the uplink will increase the power consumption of the UE even further and raise the uplink interference level of the network. Normally, those UEs which are close to the site are more likely granted 16QAM by the system, the extra noise rise created by those users using 16QAM may cause quality degradations to other users in the cell, especially those at the cell edge.

Adding higher order modulation increases the peak throughput of a UMTS/HSPA system. However, since this feature requires much higher SNR, the types of radio environments which meet the criteria are limited. For operators, it is important to understand that these new features (MIMO and HOM) are basically focussed on improving the user peak rates. Since they provide modest capacity gains, the spectral efficiency of HSPA still relics on features such as advanced receivers which improve the radio link performance.

8.2.3 Advanced Receivers

8.2.3.1 Receive Diversity and Equalizer

Rake receivers are the most commonly used front end architecture in a CDMA system. For most Rel.'99 channels which have high spreading factors (128 or 256), Rake receivers in general give very good performance. However, due to the relatively low spreading factor of HSDPA (16) compared to voice radio bearers, the lower spreading gain makes more it susceptible to poor performance in channels with large delay spreads. The inability of the Rake receiver to resolve the multipath results in an increased interference level and error rate on the data channels.

Receive diversity (Rx diversity) and equalization have been proven to be effective in enhancing the HSDPA performance. Realizing the limitation of the conventional Rake receiver based implementations, 3GPP sets the performance expectations for three types of enhanced receivers:

- Type 1: receiver with receive diversity;
- Type 2: receiver with chip level equalizer; and
- Type 3: receiver with receiver diversity and chip level equalizer.

Type-1 and Type-2 receivers have been implemented in many UE vendors' data card solutions and have shown significantly improved data performance. The Type-1 receiver provides significant improvement in HSDPA performance, especially in areas with low SNR and geometry. The Type-2 receiver, with the chip level equalizer, is capable of cancelling the own-cell interference and provides performance improvement in high geometry environment. The combination of the receive diversity and chip level equalizer (Type-3) leverages the performance gain of both of the techniques and provides a robust receiver which delivers performance enhancements for both low and high geometry environments. The utilization of advanced receivers will result in increased cell data capacity, with gains of up to 60% with Rx diversity (Type-1), 40% with Equalizer (Type-2) and 80% for the combined Type-3, compared to a sector using regular Rake receivers [5].

8.2.3.2 DL Interference Cancellation

Enhanced performance requirements for Type-3 receivers (referred as Type-3i) based on receive diversity and LMMSE (Linear Minimum Mean Squared Error) equalization for HSDPA UE were added to the standard between 2006 and 2007. The advanced Type-3i receiver is an interference-aware receiver which not only can suppress the self interference caused by multipath, but also can mitigate the interference from neighboring cells. To mitigate the interference from other cells, it is important to establish the proper interference model and related Dominant to Interferer Proportion (DIP) ratios.

As we know, the network geometry is defined as the ratio between the own cell power/ interference and the total interference power from other cells (Ior/Ioc). The interference from other cells can be divided into identified and unidentified sources. The identified interferers are known to the UE and therefore can be detected and cancelled by the advanced receiver. Those unidentified interference sources will not be resolved by the receiver and will be contributors to the overall system noise floor. The overall gain the advanced receiver can achieve depends on the percentage of interference which can be identified. Figure 8.5 shows the study results of the distribution of identified interference in a network. It includes both the simulation and field testing results which are in good agreement [6].

It can be seen that the probability that more than 80% of the interference can be identified is greater than 80% for the 0 dB geometry case. It is also shown in the field results that 57% of the time more than 90% of the interference is contained in the two strongest interferers.

3GPP has done extensive studies (simulations and field testing) on the relationship between the DIP values and the network geometries and their performance impacts. Table 8.1 shows simulation results of the performance gain of a Type-3i receiver over a Type-3 for different network geometries [7].

It is clear that using the actual DIP values (variable DIP case) provides better performance gain than using fixed DIP values. The overall gain of using a Type-3i receiver can be translated into the E_c/I_{or} gain. Depending on the network geometry, the peak again can be as much as 2.8 dB [7].

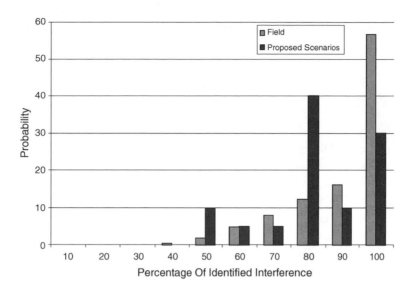

Figure 8.5 Distribution of identified interference for 0 dB geometry © 2008 3GPP

System level studies have indicated that Type-3i receiver can provide 20% to 55% coverage gain depending on the channel condition and user locations [3]. The improved link budget can also translate into system capacity gain and higher spectral efficiency for HSDPA services.

From the operator's perspective, introducing the Type-3i receiver to the network is feasible since minimum implementation complexities are involved in adding the new capability to terminal devices. Furthermore, most of the capacity gains can be realized if these type of receivers are deployed in targeted devices that are known to generate the majority of the data traffic, such as data cards or smartphone devices. No upgrades on the network side are needed.

8.2.4 Continuous Packet Connectivity (CPC)

With the deployment of HSPA, the operators' networks are becoming increasingly capable of delivering different data applications. For many wireless subscribers, making phone calls is becoming only a small portion of the daily usage of their handsets. Staying connected all the

Table 8.1 Type-3i receiver average gain over Type-3 under different geometries

	Ior/Ioc = 0 dB			Ior/Ioc = −2 dB		
	QPSK		16 QAM	QPSK		16 QAM
Ec/Ior (dB)	−6	−3	−3	−6	−3	−3
Variable DIP	23%	13%	25%	98%	19%	114%
Fixed DIP	13%	8%	14%	56%	11%	71%

© 2008 3GPP

time with the surrounding world is a concept which has been embraced by many of us today. Users who are used to slow data connections and low throughput 2G services now are exposed to many new applications (traffic on demand, video sharing, map etc.) which can only be efficiently delivered by new technologies such as HSPA. These applications in general demand constant information updates through always-on connectivity. Some of them require the transmission of real-time information with low latency. To meet such requirements, the system needs to minimize the frequency of channel switching and termination which introduce overhead signaling and delay due to the radio channel reestablishment.

In pre-Rel.'7 systems, channel switching (Cell-DCH to/from Cell-FACH to/from Cell-PCH) was implemented to keep the number of active users low. Users with low activity were downgraded to Cell-FACH or Cell-PCH so that other users could get into the system. The idea was to save system radio resources in order to increase the number of 'active' users the system could support. However, the drawbacks were an increased delay due to channel re-establishment and higher noise rise due to excessive signaling on the control channels since only open loop power control is implemented at the pre-establishment stage. For applications such as VoIP, online chat or online gaming with small packets and constant gaps between transmissions, there will be many channel switching, termination and re-establishment events within one application session. The increased application latency forces the network to spend more time and overhead resources to deal with small packet transmissions. This in turn reduces the efficiency of the network.

The limitations in Rel.'6 or older systems are the number of HSPA users supported by a cell and the latency introduced by the channel switching. On the downlink the main factor limiting the number of simultaneous HSDPA users is the code resources. On the uplink, since the spreading factors for E-DCH are low ($SF = 2$ and $SF = 4$), the mobile has to transmit more power to compensate for the loss of spreading gain. The noise rise on the uplink which in general is not a concern for a WCDMA system becomes the limiting factor to support larger numbers of active data users.

3GPP Rel.'6 already identified some of these problems and included the Fractional DPCH feature, which was really the first step towards a system that could achieve Continuous Packet Connectivity, as is discussed in the following subsection.

8.2.4.1 Fractional DPCH

Fractional dedicated channel (F-DPCH) was introduced in Rel.'6 to replace the associated dedicated channel (A-DPCH), thus reducing the signaling channel overhead. Figure 8.6 illustrates the new channel structure.

As has been discussed, each HSDPA user is associated with a dedicated control channel (A-DCH) which carries both the TPC commands and RRC signaling information. There are two main limiting factors when using A-DCH: (1) the number of HSDPA users that can be supported is limited; and (2) the power consumption of the signaling channel (A-DCH) can limit the capacity of the system. For a system with many low throughput users, the code and power consumed by the A-DCH will become the bottleneck.

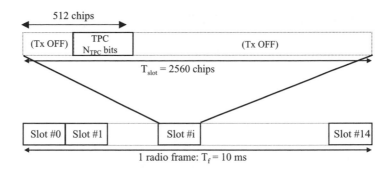

Figure 8.6   F-DPCH channel structure

F-DPCH splits the control information between the TPC bits, which are used for the fast power control on the signaling link, and the rest of the information (TFCI, pilot, signaling messages contents, etc.). While the TCP bits are still sent over the DCH channel, the remainder of the information can be sent at a higher speed tunneling it on the HS-DSCH. Therefore, the F-DPCH can be shared by multiple users in the time domain and more active users can be hosted by the system without having a big impact on the system capacity. Users with an active session, while not transmitting, only consume a small portion of the system resources and don't require re-establishment when data transfer is required. Since fewer system resources are used for signaling, the system can afford to have more users stay active in the system. This in turn avoids having unnecessary channel switching or releasing the channel within a session, and the latency caused by the re-establishment of the radio channel can be greatly reduced.

The F-DPCH technique has the following benefits:

- The hard limit of HSDPA users imposed by the code availability for their associated DCH is greatly improved. With F-DPCH, up to 10 users can be multiplexed in the same physical DPCH.
- There is an improved user experience due to the possibility of staying longer in the cell DCH state.
- The improved signaling time achieved by sending the control messages at a higher speed, which can have a significant effect on the performance of cell changes in HSDPA.

In summary, with F-DPCH the users' perception of the network quality will be improved due to the reduced application response time. Furthermore, the utilization of F-DPCH will improve system capacity due to the more efficient signaling scheme. In particular, using F-DPCH on HSDPA can provide 15% more VoIP capacity comparing with using A-DCH [8].

Since the introduction of F-DCPH Rel.'6, 3GPP added several new features in Rel.'7 to further address Continuous Packet Connectivity for HSDPA data services. The followings are the main work items in Rel.'7:

- uplink discontinuous transmission (UL DTX);
- CQI reporting reduction;
- downlink discontinuous reception (DL DRX);
- HS-SCCH-less operation; and
- new DPCCH slot format.

These features are described in the following subsections.

8.2.4.2 Uplink Discontinuous Transmission (UL DTX)

The continuous transmission of the uplink DPCCH is a major factor of the uplink noise rise. The uplink DPCCH keeps the UE synchronized when it is not transmitting. It also carries the TPC information. With an increased number of active HSDPA users in the cell, the impact of the uplink DPCCH to the system capacity becomes significant since all those channels will be active as long as the UE is not dropped from the Cell-DCH state, even when there is no application data transfer on the HS-DSCH. The interference created by these channels will not only have an impact on the data performance, but also degrade the voice quality and reduce the overall cell capacity. 3GPP studies show that the capacity loss caused by inactive users in Cell-DCH is substantial.

The goal of the uplink discontinuous transmission (UL DTX/gating) is to reduce the DPCCH transmission by turning it off when the UE doesn't have data in the buffer to transfer during the application session (e.g. as might be typical in web browsing, VoIP, online gaming, etc.). Figure 8.7 shows one of these study cases for a Pedestrian A channel based on the simulation conditions defined in Table 8.2. As can be seen, uplink gating allows the cell to maintain the throughput while keeping more voice users on Cell-DCH, or vice versa, increase the throughput with the same number of voice calls. With 25 voice calls, the UL throughput can

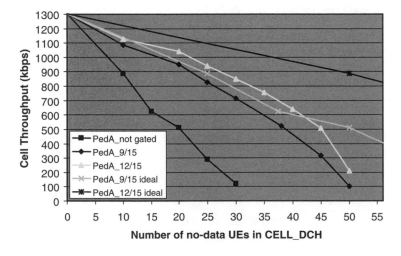

Figure 8.7   Cell throughput vs. number of inactive users in Cell-DCH [9] © 2008 3GPP

Table 8.2   Simulation assumptions for uplink gating

UE speed	Pedestrian A 3 km/hr	6 full buffer users, rest with no data
Traffic model	Full buffer, no data	
E-DCH Bitrate	{64, 128, 256, 384, 512, 1024} kbps	
DPCCH CIR target	−18.1 dB	
Load target	6 dB	
DPCCH gating pattern for no-data users	{12, 9, 0} slots gated in every radio frame.	0 gated slots equals to no gating
		Full buffer users transmit contunuously

© 2008 3GPP

be increased from 300 kbps to 800 kbps [9]. The first result (no gating) is in line with our field measurements for HSUPA shown in Chapter 6.

In general, during any packet session when the user is transmitting data in the uplink, the DPCCH is continuously active as long as the data or HS-DPCCH transmission is taking place. When there are no activities on the uplink, the DPCCH gating pattern would be applied reducing the consumed uplink capacity to a fraction compared to continuous DPCCH. By reducing the unnecessary transmission on the uplink, the interference from the HSDPA control channels will be greatly decreased. In addition, since the UE spends less time transmitting, the battery consumption will be lower. Figure 8.8 is an example showing how uplink gating works [9].

The capacity gain of uplink gating depends on the data traffic pattern on the uplink and the channel switch timer settings for Cell-DCH. For small packet applications, to mitigate the uplink noise rise and increased latency caused by constant signaling, it is recommended that the system keep the users in an active state for longer periods of time. Therefore, the benefits of uplink gating for such applications will be the improved latency without negatively impacting the overall uplink capacity. For applications with high activity factors such as FTP, the gain will be lower since most of the time the regular DPCCH will be used. Figure 8.9 depicts a 3GPP simulation study on the impact of uplink gating to VoIP capacity [9] showing the capacity gain with the gating mechanism at velocities of 3 km/hr and 100 km/hr. The DPCCH update interval is set to every eighth TTI. A 3-slot DPCCH preamble is used before actual data is transmitted. The relative capacity gain with gating is in the order of 80% for 3 km/hr and in the order of 70% for 100 km/hr.

Uplink gating, combined with DL DRX is also an important feature to improve the battery life of PS services. A VoIP call using DTX and DRX can save around 70% of the battery life compared to when these features are not used.

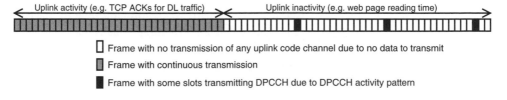

Figure 8.8   Uplink data transmit pattern with gating [9] © 2008 3GPP

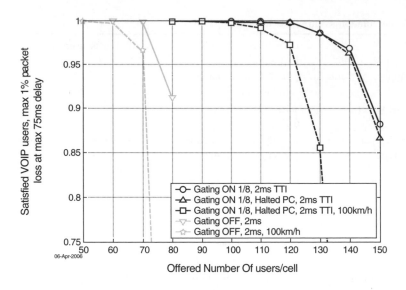

Figure 8.9   VoIP capacity gain with uplink gating [9] © 2008 3GPP

8.2.4.3 CQI Reporting Reduction

CQI reporting reduction is another important feature aimed at reducing the uplink interference. Instead of transmitting the CQI information on the HS-DPCCH all the time, the UE only transmits the CQI to the Node-B when they match the uplink DPCCH burst in the burst pattern (uplink gating). When the HSDPA user is in active transfer mode, the UE will follow the normal CQI reporting cycle which is configured by the network and sent to the UE. The critical part of this feature is that the RNC is not involved in the decision making since the RRC signaling takes longer. The gain of CQI reporting reduction depends on the initial frequency of the CQI reporting cycle. The higher the frequency, then the higher the gain is.

8.2.4.4 Downlink Discontinuous Reception (DL DRX)

Downlink discontinuous reception (DL DRX) allows the UE to switch off the receiver after a certain period when the HSDPA user has no activity by working together with the uplink gating feature. The major benefit of this feature is lower UE battery consumption.

8.2.4.5 HS-SCCH-Less Operation

As presented in Chapter 2, the HS-SCCH channel is used to carry control information that helps the users decode the HS-DSCH channel properly. Such information includes modulation and transport block format, and target UE identity. The number of HS-SCCHs determines the maximum number of users that can be multiplexed in a single TTI; in theory this number could be 15, but practical implementations reduce it to 3 or 4 because this channel consumes DL power which results in a waste of capacity.

In the case of multiple subscribers using applications with small packet transfers, it would be beneficial to multiplex a large number of these users in a single TTI, however the existing implementation of the HS-SCCH would present a capacity problem. It is not efficient to use HS-SCCH to serve multiple users with small packet applications such as VoIP in which the traffic resource utilization is equivalent, or even lower, than that of the signaling overhead.

To solve this problem, HS-SCCH-less operation was introduced. With this feature it is possible to transmit data without the need for the HS control channel. As mentioned before, this feature was mainly introduced for small packet applications such as VoIP. In VoIP applications, the traffic pattern is fixed, predictable and delay sensitive. All users have the same scheduling priority.

In HS-SCCH-less operation, the first transmission always selects QPSK and only four pre-defined transport formats (TF) are used for the UE to blindly detect the correct format. The four possible transport formats are configured by higher layers. Predefined channelization codes are used for this operation mode and are configured per UE by the higher layers: the parameter HS-PDSCH code index provides the index of the first HS-PDSCH code to use. For each of the transport formats, it is configured whether one or two channelization codes are required. There is no HS-SCCH associated with the first transmission needed since many of the information have been predefined as shown in Table 8.3.

The benefits provided by HS-SCCH-less operation are:

- improved downlink capacity, especially for real time services such as VoIP;
- reduced HS-SCCH usage on VoIP allows more best-effort traffic on HSDPA; and
- delay sensitive traffic is better supported as more users can be code multiplexed into a single TTI without the costly overhead of the HS-SCCH.

As shown in Figure 8.10, for the same best effort throughput target, the number of 12.2 kbps VoIP users supported by the cell can increase by up to 50% [9].

8.2.4.6 New DPCCH Slot Format

With the introduction of uplink gating, the Rel.'6 DPCCH channel format, which is more suitable for the case when data is transmitted on the HS-DSCH channel, is not necessarily the

Table 8.3 HS-SCCH information which are not needed in HS-SCCH-less operation

Information in regular HS-SCCH	HS-SCCH-less operation
Modulation information (QPSK or 16QAM)	QPSK only (no signaling need)
OVSF code information	Predefined one code
TF indicator	Blindly detected by UE (only two TF)
HARQ process number	No need for synchronized IR
Redundancy version information	Predefined
New data?	No need for synchronized IR
UE identity	No need, carried in the HS-PDSCH

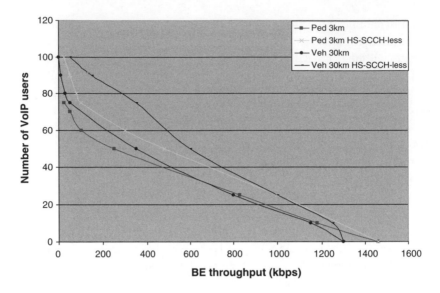

Figure 8.10   HS-SCCH-less capacity gain for VoIP and BE mixed service © 2008 3GPP

one which can minimize the overhead when DPCCH is the only uplink channel transmitting. A regular DPCCH channel carries HARQ acknowledgment and CQI reporting information. Within each slot, there are five to eight pilot bits to provide reference for channel estimation for data decoding and two Transmit Power Control (TPC) bits for power control. There are several new DPCCH slot formats proposed in the standard. In general, the idea is to redistribute the pilot bits and TPC bits in the slot. The achieved E_b/N_0 gain for the same TPC error rate is about 2 to 3 dB. Figure 8.11 shows the case for PA3 [9].

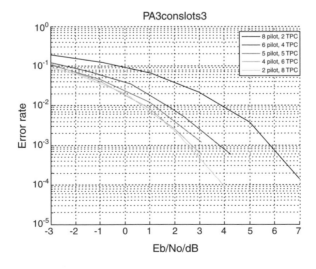

Figure 8.11   TPC error rate for different new DPCCH slot format © 2008 3GPP

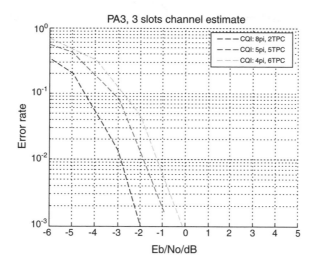

Figure 8.12 CQI report error rate for different DPCCH slot format © 2008 3GPP

Basically, the benefit of the new format is that the DPCCH can transmit at lower power through the SIR target reduction for TPC bits, which causes less interference on the uplink. However, since the reduction of the pilot bit is at the cost of the channel estimation, the performance of the HS-DPCCH may be degraded. The error rate for CQI and HARQ acknowledgment could increase as indicated by the simulation results shown in Figure 8.12 [9]. A 2 dB degradation can be expected if the new format is used for the CQI reporting. For HARQ acknowledgment, the impact is less severe.

Given these pros and cons, it is recommended that the new DPCCH slot format be implemented together with SIR target reduction, CQI reporting reduction and uplink gating to provide maximum benefit on the uplink capacity improvement.

8.2.5 Circuit-switched Voice Over HSPA

Even though wireless VoIP is a hot research topic these days, there are still a lot of hurdles to overcome for operators to implement it on a mass scale. The slow adoption of IMS and the lack of support of Voice Call Continuity (VCC) present big barriers for operators who are expecting a smooth transition when introducing VoIP service to their customers. In the past decades, circuit-switched voice has set the standard for voice service quality and reliability, so packet-switched VoIP must meet or exceed those expectations. In addition, to achieve the full benefits provided by VoIP, an efficient all-IP packet data network with low latency is needed. This in turn requires substantial network architecture changes which may not be very appealing to operators who just deployed UMTS/HSPA networks.

As an alternative to improve the spectral efficiency by utilizing the existing infrastructure, Rel.'8 standards introduced the CS voice over HSPA feature. The general idea is to use the

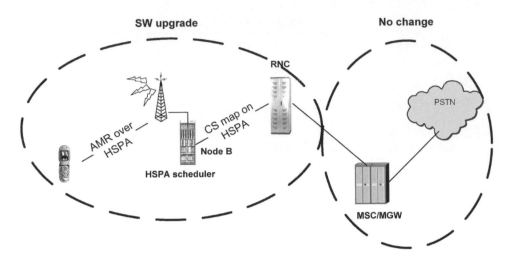

Figure 8.13   CS voice over HSPA

existing CS core structure, therefore from the UE and core perspective it is still a circuit
switched voice call. The only difference is that the voice data stream is tunneled through HSPA
as shown in Figure 8.13.

CSoHS is a promising feature that is likely to be deployed as a first step towards a converged
'All-IP' network. There are many benefits provided by this feature:

- No significant impact on the infrastructure. No core network upgrade is needed. Software
  upgrade is needed on RAN and UE.
- Improved spectral efficiency by moving Rel.'99 traffic to HSPA.
- Existing CS mobility procedure can be applied for simple VCC implementation.

8.2.6 Dual Carrier Operation in HSDPA

This feature is a work item in Rel.'8. The concept of carrier aggregation is similar to that of
multi-carrier (MC) CDMA approaches. The idea is to aggregate multiple 5 MHz UMTS
carriers together to create a bigger pipe of 10 MHz in the downlink; there is no change to the
uplink transmission, which will still operate over a 5 MHz bandwidth. Since multiple carriers
can be allocated to the same user, higher throughput and lower latency can be expected.
Because the aggregated carriers are jointly scheduled, the capacity gain is beyond linear; that is,
the capacity gain of carrier aggregation will be more than adding independent multi-carriers
since they benefit from the additional trunking efficiency associated with the larger 'channel
pool'.

The benefits of this feature are improved data throughout, latency performance, and better
spectral efficiency compared with independent multi-carrier deployment for HSPA. However,

the drawbacks are obvious as well: it requires at least 10 MHz of adjacent spectrum to be allocated to HSPA to achieve these gains. This could be a limiting factor in many deployment scenarios.

8.3 Architecture Evolution

While the radio improvements for HSDPA and HSUPA have had a steady impact on the performance and efficiency of the air interface, important changes have also been standardized in the overall network architecture that will in some cases complement the features on the radio side, or add on to the final performance improvement. In this section, we focus on two of the major items standardized during Rel.'7 and Rel.'8:

- GPRS flat architecture.
- End-to-end QoS architecture.

The first item is an initial step approach to a flat architecture for the GPRS core, in which the user plane of the data services is directly routed from the RNC to the GGSN, bypassing the SGSN. This can result in higher backhaul efficiency and reduced end-to-end latency, not to mention the savings this could represent to the operator in terms of SGSN capacity. The next section will provide further details on this particular feature.

The latter part of this section presents the new 3GPP QoS architecture as defined by Rel.'7 and Rel.'8. As explained in Chapter 3, a fully consistent end-to-end QoS architecture is a very important functionality to deploy in multi-service wireless data networks, because it increases the efficiency of the network.

End-to-end QoS architecture standardization, although already started on Rel.'6, is not mature until the improvements brought in Rel.'7. This new architecture represents a good scheme to control the quality of the applications not only on the radio side, but in all instances within the operator's network, with a high level of interworking between 3GPP and external items, such as elements from the Internet Engineering Task Force (IETF) world. Section 8.3.2 covers the architectural changes and what these represent to operators who expect to heavily deploy data services.

8.3.1 GPRS Flat Architecture

The evolution of the GPRS core towards a flat architecture provides two main benefits:

- *Latency decrease* on the overall end-to-end transmission path, and
- *Cost saving*, scalable solution for high-volume data operation.

It has previously been discussed how the efficiency and performance of data services has been greatly enhanced with subsequent standardization improvements after UMTS was initially deployed: Rel.'5 introduced HSDPA, Rel.'6 HSUPA and the new Rel.'7 and Rel.'8

include improvements to both technologies that will further increase spectral efficiency, reduce network latency or improve peak throughputs. Although these features are good, they all improve the Radio Access Network while leaving the core network as is. Furthermore, the core network architecture has remained essentially untouched ever since it was first designed to carry the first GPRS data traffic in 1997. As is explained in the next section, such a legacy core architecture is quite inefficient for transmission of high data volumes and high data rates.

Next generation wireless networks, such as WiMAX and LTE, apart from being evolved radio interfaces, agree upon the concept of a simple core network architecture: a PS-only core with a reduced number of nodes (just one type of transit node). The advantage is minimizing latency and savings on CapEx and OpEx. The evolution towards a GPRS flat architecture in HSPA is a natural step to keep the system competitive and extend its life well into the future.

Initial attempts to improve the GPRS architecture started as early as 2000, with proposals such as Nokia's distributed All-IP RAN that even underwent product trials in 2002. However, the initiative may have been too advanced for its time and there was insufficient market demand for it. In recent years this debate has been restarted, boosted by the new architecture models proposed by both WiMAX and LTE; several architectural changes have been proposed in 3GPP. The first of these changes has already been fully standardized in Rel.'7 and is known as 'Direct Tunnel' functionality [10,11].

8.3.1.1 Rel'7 Direct Tunnel Functionality

In the current GPRS architecture, the PS functionality is distributed in three main logical units: RAN, SGSN and GGSN, as shown in Figure 8.14.

The UMTS Radio Access Part (UTRAN) takes care of the radio transmission and reception, including segmentation, error protection and general radio resource management, among other things. The Serving GPRS Support Node (SGSN) performs PS mobility management, packet routing and transfer, authentication and charging. The Gateway GPRS Support Node's (GGSN) main functionality is the routing of packets to and from external networks. It also performs authentication and charging functions, and is the main network element responsible for Quality of Service (QoS) negotiation.

Any data communication in an HSPA network begins with the establishment of a Packet Data Protocol (PDP) Context between the UE and the GGSN. The data corresponding to each specific PDP context is first carried by the GPRS Tunneling protocol (GTP), and then by the Packed Data Convergence Protocol (PDCP) layer. These protocol layers create a 'pipe' for transfer of IP data from the UE to the GGSN, hiding all the underlying complexity related to radio networking, as can be observed in Figure 8.15. Therefore, once the tunnel is created, the traffic effectively flows from UE to GGSN, with minimal modification in the intermediate nodes apart from the effect of the lower-layer radio protocols.

While the initial GPRS architecture has been sufficient for the amount of traffic and the performance requirements for 2G and early 3G systems, it presents some challenges in a more advanced data network. The fact that all the data traffic needs to be routed first to the RNC, then

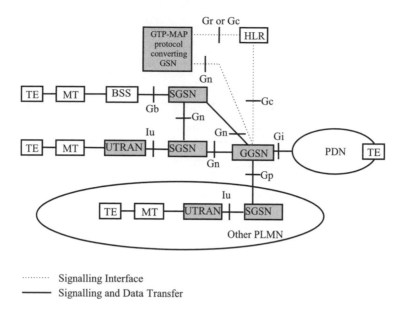

Figure 8.14 GPRS core architecture with main interfaces [12] © 2008 3GPP

to the SGSN, and then on to the GGSN represents a waste of processing power on the intermediate nodes, not to mention the increase in the latency.

The Direct Tunnel Functionality, also known as 'GPRS One Tunnel Solution', proposes a simplification of the existing GPRS core architecture for the user plane traffic, eliminating the transit of the data through the SGSN. Most of the transmitted data flows directly from the RNC

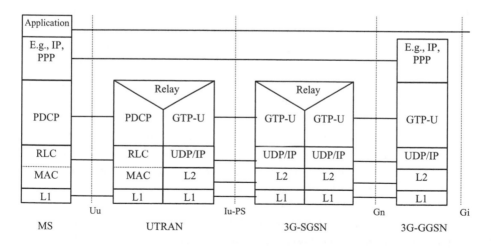

Figure 8.15 GPRS Protocol architecture in UMTS [11] © 2008 3GPP

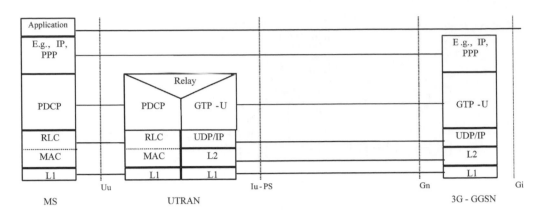

Figure 8.16   GPRS protocol architecture with Direct Tunnel [11] © 2008 3GPP

to the GGSN, eliminating unnecessary routing delays on the SGSN, and also reducing possible capacity bottlenecks from the data path when preparing the network for heavy data utilization. The control plane traffic is still routed through the SGSN, since the node still keeps its original functionality regarding mobility management, charging and authentication. Figure 8.16 shows the new protocol architecture for the user plane [11], which is notably simpler than Figure 8.15.

Since the SGSN retains the charging functionality, the standardized Direct Tunnel solution has some limitations in scenarios such as inter-operator roaming or prepaid data charging. In such cases, the previous architecture of Figure 8.15 will be applied.

8.3.1.2 Further Architecture Standardization (Rel.'8) Beyond

While the Direct Tunnel solution improves the efficiency of the GPRS architecture, it does not achieve a fully flat architecture because the packets still need to be routed through an intermediate node, the RNC. While the RNC has important radio control functionality for Rel.'99 channels, in the case of HSPA channels most of this functionality has been pushed down to the Node-B, therefore managing all this traffic at the RNC level results in an inefficient utilization of resources. This situation was identified and discussed in 3GPP, where Rel.'7 includes several architectural alternatives to tackle the problem [13]. The recommended solution (alternative 4) merges RNC functionality with the Node-B on the basis that none of the existing interfaces would be modified. The specification defines two different deployment scenarios, one with PS only services (stand-alone) and one scenario with mixed CS and PS traffic (carrier-sharing scenario). Figure 8.17 illustrates the concept.

As the figure illustrates, evolved HSPA Node-Bs have a direct IP broadband connection towards the GGSN for PS traffic. This further optimizes the latency and facilitates the network scalability without capacity bottlenecks.

With the evolved HSPA architecture, the Iub link is replaced by the Iu-CS, Iu-PS and Iur. The Iu-CS and Iu-PS links are split between user plane and control plane, each of them going to a separate entity:

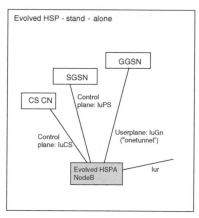
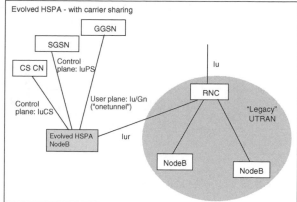

Figure 8.17 Evolved HSPA Architecture [13] © 2008 3GPP

- the Iu-PS interface (control plane) is routed to the SGSN;
- the Iu-PS interface (user plane) is routed to the GGSN, as in the Direct Tunnel case;
- the Iu-CS interface (control plane) is routed to the CS core; and
- the user plane for CS traffic is routed to the RNC via the Iur link.

While the implementation of such a solution doesn't require any modification of the existing standards, Rel.'7 and Rel.'8 specifications introduce some special protocol support to enable smooth operation when deploying this architecture. On the other hand, the deployment of this architecture is completely optional and independent from the implementation of the radio enhancements provisioned by HSPA+.

Having the RNC functionality for PS embedded in the Node-B poses some challenges to the treatment of mobility situations. In particular, implementing soft-handover combining in uplink is practically impossible due to the excessive amount of information that should be sent through the Iur link, therefore UL will not benefit from SHO gain in the border between cells. Also, cell transitions in downlink will be more abrupt because in practice the connection goes through an 'inter-RNC' HS-DSCH cell change procedure, which requires SRNS relocation. As has been demonstrated in prior chapters, this interruption is larger than in the normal intra-RNC case; however infrastructure vendors believe it can be implemented with a very small interruption (below 200 ms), otherwise this architecture could not be used to carry real-time or streaming services.

8.3.1.3 Benefits from the HSPA Evolved Architecture

On the benefits side, having a flat architecture translates into a performance improvement on data services due to the reduction in latency. This is illustrated in Figure 8.18, which shows the latency improvement measured as the Round Trip Time (RTT) for a 32 byte packet, for

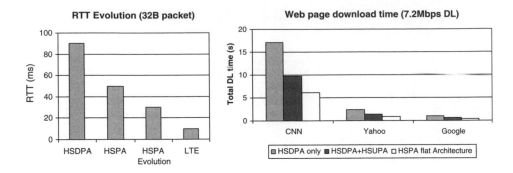

Figure 8.18 Improvement of RTT with HSPA Evolved architecture (left) and impact on web performance (right)

different air interface technologies (left). The chart on the right demonstrates how an improved latency is reflected in the end-user performance, in this case, for download times of different web pages. The results shown are just illustrative examples based on a web page download analytical model.

Having a very small latency also plays an important role when offering *real-time services* in the network. Real-time services have very stringent end-to-end delay requirements, and the network latency ultimately affects the system capacity to carry this type of services. Since network resources are limited, there is a tradeoff between achievable delay and load level in the network: the more loaded the network is the more difficult it is to achieve a low end-to-end delay. Having shorter transport latency permits the network to operate at higher load levels and achieve better resource utilization efficiency.

Another clear benefit of the HSPA Evolved architecture is the scalability and possible cost savings of the solution for high volumes of HSPA traffic, since growth in data traffic will not require capacity extensions in the RNCs. The reduction in RNC equipment can be achieved even when only a small percentage of the deployed nodes are HSPA Evolved Node-Bs. Figure 8.19 illustrates a simplified deployment scenario for a medium-size operator (30 000 cell sites) after initiation of HSPA service offering. The analysis assumes that 10% of the sites are hotspots, i.e., they carry intensive data traffic, while the rest of the sites have lower data demand. It has been assumed that the peak rate on the hotspots were 0.5, 1, 3, 6 and 9 Mbps during the five-year period, while the rest of the nodes experienced moderate growth (20% yearly increase). In this calculation it was assumed that an RNC can support up to 1000 MB/s in the Iub interfaces.

The analysis shows that by the fifth year, having a hotspot deployment with HSPA Evolved architecture could save nearly 70% of the total RNC investment for data services. Further RNC savings could be possible if the voice traffic were offered through the VoIP service, which does not require CS infrastructure and can be treated as data traffic as well. The final section of this chapter will provide further analysis on the Voice over IP technology for HSPA networks.

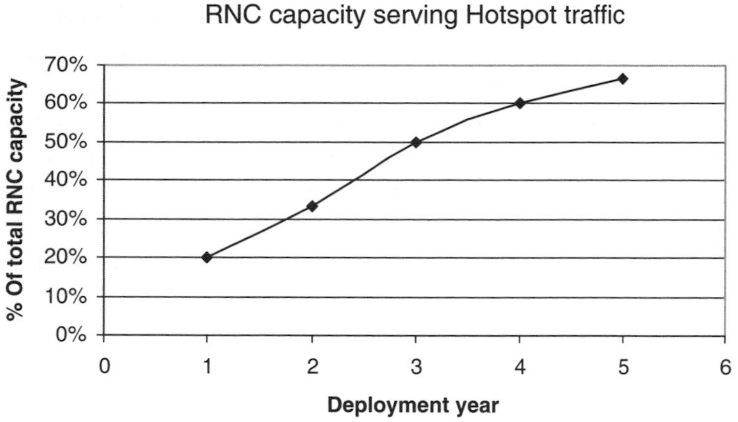

RNC capacity serving Hotspot traffic

Figure 8.19   RNC capacity savings in a hotspot deployment scenario

8.3.2 End-to-end Quality of Service (QoS) Architecture

As discussed in Chapter 3, when wireless data networks reach maturity – in terms of traffic volume and multi-service environment – the treatment of traffic through QoS mechanisms becomes more and more important. While 3GPP has introduced QoS treatment in all the previous releases, it is with Rel.'7 and further Rel.'8 improvements that the QoS architecture has reached a complete end-to-end scope, including tight integration with other networks (from other operators or different technologies).

With the enhancements in Rel.'7 and Rel.'8 it is possible to fully control the service quality and charging aspects of a variety of scenarios, such as:

- Identify a heavy data user that is loading the network and impacting other users, and apply corrective actions, such as limiting its maximum throughput or preventing access to specific applications (like peer-to-peer).
- Select the best wireless access at a given geographical area (for instance UMTS or WiFi) and charge the user accordingly.
- Negotiate QoS parameters for an end-to-end communication that terminates in a different network operator.
- Apply different charges for different services offered by the operator.
- Charge a premium for high-quality data services during peak hours.

As indicated before, QoS treatment has been an essential part of 3GPP specification work for data services from the very early releases of GPRS in Rel.'97, followed by continuous improvements in Rel.'99, Rel.'4, Rel.'5 and Rel.'6. Up to Rel.'4 all the QoS mechanisms were very much focussed on the RAN part of the network, while with Rel.'5 and Rel.'6 new elements were created to introduce some degree of control over the application level. The main new elements, called Policy Decision Points and Policy Enhancement Points, acted together to define what level of resources and QoS class could be granted to a particular application flow, based on the service type, subscriber information, operator policy, etc. Improvements in Rel.'6 also enabled a certain control over the QoS at the Application Level (end-to-end), although its scope was limited to IMS services. Unfortunately, the QoS versions prior to Rel.'7 were not practical because they didn't provide full end-to-end support, including the coordination with other network elements not specified by 3GPP such as IP routers, and very few (if any) of such architectures have ever been deployed.

Rel.'7 QoS specifications, after the learning process from the previous releases, are perceived by vendors and operators to be a mature QoS architecture that can fulfill the needs of a complex and generic application environment. The main items introduced in this release are:

- improved end-to-end QoS negotiation;
- improved charging and QoS enforcement with multi-technology networks; and
- better integration with external networks.

Figure 8.20 presents a simplified scheme for the new QoS architecture, including the major elements such as the Policy and Charging Rules Function (PCRF), the Policy and Charging Enforcement Function (PCEF) and the Application Function (AF).

8.3.2.1 Improved E2E Charging and QoS Enforcement

The Policy and Charging Control (PCC) architecture was evolved to support dynamic QoS and charging policies to all network elements. The architecture was designed taking into

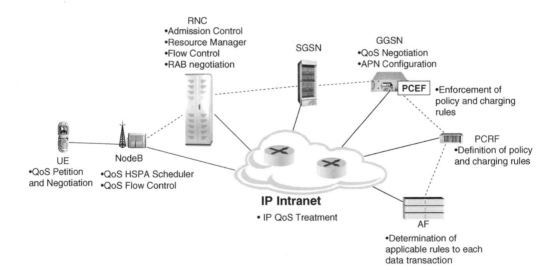

Figure 8.20   QoS architecture introduced in Rel.'7

consideration technologies other than the 3GPP family, with the intention to create a universal solution for QoS management. The PCC framework as defined in Rel.'7 can be used in multi-technology environments, where a subscriber can access multiple applications via different access technologies (for instance, 2G, 3G, WiFi), ensuring a uniform level of end-to-end QoS and facilitating charging from all the involved network elements. In order to realize these objectives, the logical elements introduced in earlier releases have been updated with new functionality and consolidated into the following units, shown in Figure 8.21:

- Policy and Charging Rules Function (PCRF);
- Policy and Charging Enforcement Function (PCEF); and
- Application Function (AF).

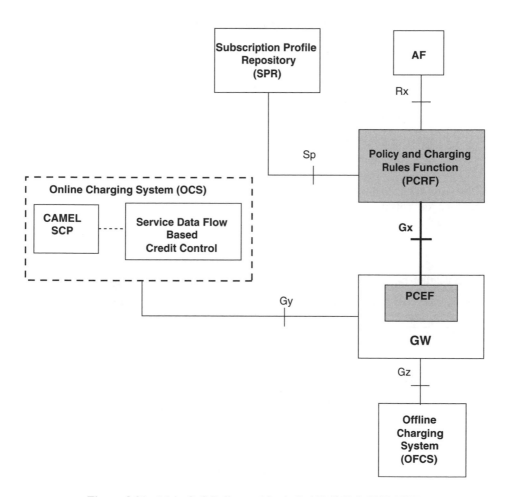

Figure 8.21   Main QoS Policy entities in Rel.'7 [14] © 2008 3GPP

The Policy and Charging Rules Function (PCRF) evolved from the Rel.'5 Policy Decision Function (PDF) unit which was limited to determining static charging rules. The PCRF provides support for dynamic policy rules for specialized services, while allowing the utilization of pre-defined policy rules that are stored in the PCEF. The PCRF determines the policy and charging rules and then applies them to the involved gateway nodes.

The Policy and Charging Enforcement Function (PCEF), located in the GGSN, is responsible for enforcing the policy and charging rules signaled by the PCRF. Other functions include the storage of dynamic and static rules, and data flow detection and charging.

The Application Function (AF) is an element offering applications to the 3GPP network. This unit needs to communicate with the PCRF in order to determine the rules to apply to a specific data communication. Such communication is maintained for the duration of the data session, as the policy rules may change.

8.3.2.2 Enhanced Integration with External Networks

The QoS interworking support has been reinforced in Rel.'7 standards. It defines improved methods to integrate with IP networks in particular, as shown in Figure 8.22. The QoS information is defined for the end-to-end communication and propagated accordingly to the different layers and elements of the connection. In addition to this, the QoS information needs to be propagated from one end (client) to the other (server) and vice versa.

The following elements are needed to ensure the end-to-end operation and a better integration with other network protocols:

- **IP BS Manager:** manage the external IP bearer service, and communicates with the UMTS BS Manager. This unit is located in the GGSN, and may optionally be implemented in the handset.

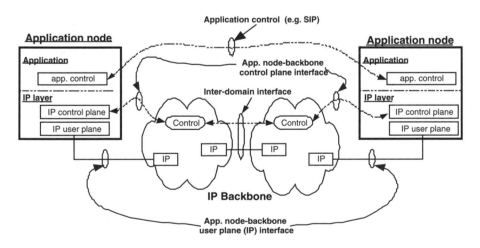

Figure 8.22   Integration of 3GPP QoS with IP QoS [15] © 2008 3GPP

Table 8.4 Example of 3GPP traffic class mapping with DiffServ

3GPP traffic class	Traffic handling priority	DiffServ code point
Conversational		Expedited Forwarding
Streaming		Assured Forwarding 1
Interactive	1	Assured Forwarding 2
Interactive	2	Assured Forwarding 3
Interactive	3	Assured Forwarding 4
Background		Best Effort Forwarding

- **Translation Function:** provides inter-working between the mechanisms and parameters used by the UMTS bearer service and the ones used by the IP bearer service.

IETF defines two ways to handle QoS in IP networks: Integrated Services (IntServ) and Differentiated Services (DiffServ). IntServ was designed inspired by the mechanisms to guarantee the quality in CS networks, reserving specific resources for each particular traffic flow. On the other hand, DiffServ is not focussed on traffic flows but on traffic classes and provides a coarser, although more scalable differentiation than IntServ.

In 3GPP Rel.'7 the support of DiffServ translation is mandatory, while IntServ is optional. With DiffServ up to 64 different traffic classes can be defined, but typically only the following are used:

- *Default* – which is typically best-effort traffic;
- *Expedited Forwarding (EF)* – for low-loss, low-latency traffic;
- *Assured Forwarding (AF)* – allows the operator to provide assurance of delivery as long as the traffic does not exceed some subscribed rate;
- *Class Selector PHBs* – which are defined to maintain backward compatibility with the IP Precedence field.

Table 8.4 provides a translation example between 3GPP traffic classes and IP DiffServ parameters.

In summary, the new QoS architecture provides the operator with multiple ways to control the applications being offered by the network, including resource control and charging aspects. The interaction with external networks ensures that the negotiated quality is maintained for elements that do not belong to the 3GPP world, and with operators with whom there is a previous Service Level Agreement (SLA).

8.4 Converged Voice and Data Networks: VoIP

As we discussed in the beginning of the chapter, the communication industry is migrating towards an 'All-IP' environment. The support of Voice over IP (VoIP) is one important piece of the puzzle in the All-IP transition, because it permits unification of the core network into a single, packet data core. Furthermore, offering the voice service over IP in an HSPA network has other additional benefits:

- **Possibility of new and richer services**. The fact that voice is carried over a data bearer instead of a CS bearer reduces the complexity of new services that mix data content with real-time communication. Applications that are successful in the internet world can be directly imported into the wireless world with minimal complexity.
- **Cost savings in the RAN**. Due to the increased spectral efficiency of VoIP over typical CS voice, and to other revolutionary features being introduced in the PS side of the network, there can be a significant reduction in the cost of the network.

VoIP is offered commercially in many landline networks with providers such as Vonage, Skype, etc. The wireless domain introduces a number of potential problems for VoIP service quality such as packet loss, delays, jitter and other effects that can be far worse and much more variable than landline networks, therefore the adoption of VoIP in UMTS/HSPA networks is not straightforward and can present many challenges. This section analyzes the possibility of deploying the VoIP service in a HSPA network, through an evaluation of performance and feasibility, identifying the benefits and the requirements to deploy it as a viable service.

Section 8.4.1 analyzes the gains of deploying an All-IP network, for which VoIP is the cornerstone. In Section 8.4.2 some fundamental VoIP concepts are introduced, providing the necessary tools to understand the challenges posed by the technology, as well as new opportunities for the operator. Sections 8.4.3 and 8.4.4 analyze the requirements to deploy a viable service, both from a service perspective, and from a performance and capacity perspective, respectively. Section 8.4.5 presents a preliminary assessment of the technology in a lab environment.

8.4.1 Benefits of an All-IP Network

As a whole the telecommunications industry is transitioning from hybrid networks (mixed IP and TDM networks) to 'All-IP' networks. The mobile sector of the telecommunications industry is no different, however, due to the more complex systems inherent in mobile networks this transition has taken longer to begin realization. In this section we review the benefits that an All-IP network can have for a wireless operator.

8.4.1.1 Better Equipment Interoperability

One of the major issues facing mobile operators is the interoperability of systems: the complexity of the networks has caused a proliferation of interface types and signaling systems. Each different interface and signaling system requires coordination and in some cases standardization in order that the devices at each end will operate with each other. If these devices are obtained from different equipment vendors interoperability testing will be required; this will need to be done by the operator or the vendor, but either approach increases costs and the time taken to introduce new services to the market. Specialized interfaces also require increased staff overheads as specialist staff are required to maintain, operate, design, and grow each interface. An 'All-IP' network reduces all of these issues by simplifying the underlying transport mechanism, and by reducing the number of signaling systems required.

One factor not seen yet but that will come into play in increasing measure over the coming years is the increasing proliferation of IP technologies in the PSTN network; either through the adoption of the technology by traditional Local Exchange Carriers (LECs), or – in more recent years – new entrants to the market offering IP phone service to consumers. If these trends continue TDM based fixed line networks will diminish in the market, or even if these trends slow down, the interconnection between carriers will transition from TDM to IP based technologies over the next few years.

8.4.1.2 Better Scalability

An All-IP network increases the overall flexibility of the network, by allowing systems to be re-purposed to account for unforeseen loading patterns or to allow new functions to be added earlier. Typically All-IP architectures separate control and bearer traffic, allowing for the flexible locating of control infrastructure, and improving the scalability of the system as a whole.

8.4.1.3 Simplified and Cheaper Transport

Existing hybrid networks need to operate multiple transport networks for access, backhaul and for core network communications. The All-IP network can utilize a single network, thereby permitting the operator to focus on optimizing one network for performance, reliability, and efficiency. A single network also improves system latency, since control messages have to pass through fewer layers, and bearer traffic has to pass through fewer nodes.

Also, the proliferation of IP hardware has resulted in a commoditized marketplace for equipment; this means that even though a telecom operator must use robust systems, the cost of this hardware is lower than similar traditional hardware. Furthermore, because of the volumes of systems developed using IP technologies, best practices are identified quicker and systemic faults corrected sooner than would be the case in traditional networks.

8.4.1.4 Simplified Core Network

The fact that both voice and packet data services can be processed with the same equipment unit provides an additional cost savings benefit to the operator, who doesn't need to maintain separate networks for CS and PS data services. This facilitates the sharing of common resources, thus improving the efficiency and scalability of the unified core network.

All of these factors contribute to the fact that an All-IP network will be cheaper to build, grow, and operate than a similar hybrid network, while simultaneously reducing the time to market of new services and functionality.

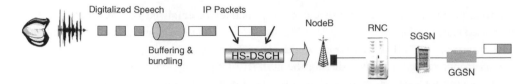

Figure 8.23 Illustration of VoIP packet communication in HSPA

8.4.2 Fundamentals of Voice over IP (VoIP)

This section introduces fundamental concepts about the VoIP service:

- explaining how voice is coded and sent through the packet network;
- defining the control protocols needed; and
- identifying the factors that can affect the performance of the service.

Figure 8.23 illustrates the concept of VoIP in a HSPA network. As is the case with the current CS voice service, the analog voice will be sampled and digitalized, and later mapped into logical and physical channels. In the case of VoIP, these channels are different than in the normal case of voice, since they will be utilizing the HS-DSCH, which is a PS shared channel as opposed to the dedicated channel used by AMR voice services. Furthermore, the digital voice is encapsulated into IP packets which circulate across the PS data network as any other data packet, subject to congestion, delays and packet loss. Normally there will be neither dedicated resources nor establishment of a communication path, since VoIP is usually conveyed over a connectionless IP service.

8.4.2.1 Voice Coding

The voice will be sampled at a sampling frequency sufficient to capture the information in the signal, typically 8 or 16 kHz, generating samples every 0.125 ms or 0.062 ms, respectively. These voice samples can be digitalized through several different mechanisms depending on the codec used. The most basic vocoders are based on Pulse Code Modulation (PCM), which quantize the voice waveform into a pre-defined series of quantization levels, each of them representing a specific amplitude level. These techniques can offer very good sound quality if the number of levels is large enough, and the sampling frequency is also fast enough, and for this reason this is the method used for digitalization of HiFi music, movies, etc. Modern HiFi sound systems such as CD, DVD and Blu-Ray are based on a variant of PCM called Linear Pulse Code Modulation (LPCM). The main drawback of these systems is the amount of bandwidth required to transmit the sound stream: for instance a typical DVD sound channel is transmitted at 768 kbps.

Due to bandwidth considerations most of the cellular systems operating today use a variation of a coding technique called 'code-excited linear prediction' (CELP). CELP codecs try to estimate the characteristics of the voice and encode it into different parameters. Such parameters are applied at the receiver and the sound sample is recreated, rather than decoded, using a speech production model. GSM EFR and AMR codecs are based on a variant of CELP called 'Algebraic CELP' or ACELP, while CDMA1X's uses RCELP (Relaxed CELP) for their EVRC codecs, and eX-CELP (Extended CELP) for their SMV vocoders. CELP codecs used in wireless can provide acceptable voice quality, or even better than landline if Wideband coding is used. Their main advantages are their robustness to bit errors – a critical factor in wireless environments – and their transmission efficiency, since they can transmit voice channels with speeds ranging between 2 and 24 kbps, far less than would be required for PCM.

While CELP codecs have proven very good with CS cellular services, in the particular case of VoIP there is a new type of codec that can offer a significant advantage over these because they have been specially designed to be resistant to IP packet loss. Cellular CELP codecs have been designed to be resistant to bit errors, but not so much to frame errors, since their coding is based on exploiting the interdependencies present between consecutive speech segments. For this reason, the performance of these codecs is very much dependent on the results from the decoding of the previous frames, and a single frame error can result in decoding problems of subsequent speech frames. Since the VoIP packets can be subject to packet loss due to congestion and specially a high delay variation, which will indirectly increase the packet loss, it is desirable to have vocoders in which the packets are encoded independently from each other. This is the basis of the iLBC (internet Low Bitrate Codec), which has become increasingly popular in the VoIP world since its standardization by IEEE in 2004.

Table 8.5 provides a resource utilization and performance comparison between popular voice codecs, including landline (G.711), iLBC and AMR. Voice quality is measured using the Mean Opinion Score (MOS) scale, with higher scores (up to a maximum of 5) being better.

iLBC codecs present a much better response to packet errors than other typical VoIP codecs such as G.729a and G.723 [16].

One advantage of the utilization of VoIP is that the technology is independent of the codec used to carry the voice information. This can potentially enable a faster evolution of speech codecs, because the cellular layers will not be affected by a codec change. It also provides the operator with the flexibility to decide what family of codecs to use, unlike the present day where GSM operators are limited to using AMR and CDMA2000 operators to EVRC.

Table 8.5 Resource utilization comparison of popular voice codecs

Codec	Source Bitrate	Quality (MOS)	Voice payload (Bytes)	Voice payload (ms)	Codec delay (ms)	Packet per second (PPS)	Physical bitrate[1] (kbps)
G.711	64 kbps	4.1	160	20 ms	20 ms	50	87.2 kbps
G.723.1	6.3 kbps	3.9	24	30 ms	37.5 ms	34	21.9 kbps
iLBC	15.2 kbps	3.9	38	20 ms	25 ms	50	38.4 kbps
AMR	4–12.2 kbps	4.06	31	20 ms	25 ms	50	35.6 kbps

[1]Ethernet bandwidth required including 58 bytes of RTP/UDP/IP + Ethernet overhead

On the other hand, having a software codec can be an important source of delay. Today's cellular codecs are implemented in hardware within the chipsets containing the protocol stack, which is very efficient from the delay point of view. If the VoIP codecs are not implemented in the hardware, there is a possibility that the delay at coding can degrade the overall experienced quality. There are two types of delay occurring in speech coding, an algorithmic delay and a processing delay [17]. The algorithmic delay is related to the way the speech samples are bundled into different blocks, while the processing delay is related to the computational and memory requirements of the algorithm. Therefore, a very complex codec is more likely to have processing delay issues than a simpler one.

8.4.2.2 IP Transport Protocol

The nature of the voice service makes it very sensible to delays, while the data loss is not critical as long as it's contained under a threshold value (for instance, the AMR voice service can afford a 2% Frame Error Rate without significant perception on the receiver). Since low packet latency is the key to the success of VoIP, voice packets are typically encapsulated into connectionless UDP blocks. These transport protocols provide a faster means to send the information across the network; however unlike TCP they do not guarantee the correct delivery of the content or the order of arrival of the packets.

UDP alone does not provide with the tools to efficiently send and process the voice packets. A very popular choice is to use RTP (Real-time Transport Protocol) on top of UDP in order to provide services to the upper layers such as:

- payload identification;
- time stamping;
- sequence numbering; and
- delivery monitoring.

While RTP does not guarantee packets to be delivered, or to be delivered in the correct order, it provides the tools to re-order the packets at the receiving end if, as is usually the case, the packets are stored in a buffer. Such a buffer, often called the 'dejitter buffer' is there to compensate for the delay variation suffered by the IP packets flowing through different transport routes. Such a buffer artificially delays the reproduction of the sound on the receiver end, but ensures that the frames are decoded at the right moment every time, as shown in Figure 8.24.

Figure 8.24   Illustration of the effect of the dejitter buffer

The length of the buffer can be adjusted based on the jitter effects on the transport route: a longer buffer can support higher delay variations, but can impact voice quality if the introduced delay is beyond a perceptual limit. In a similar fashion, there is a tradeoff between the load levels that can be allowed on the radio interface and the overall latency introduced on the packet route, since a higher radio load will delay the VoIP packets. If the different components of the end-to-end latency can be reduced (including physical radio transmission, transport and core delay components) then load in the radio interface could be increased thus achieving higher levels of network efficiency. The delay increased on the radio side would then be treated by the dejitter buffer and the overall quality would be maintained. Simulation studies indicates that a relaxation on the delay budget can boost the network capacity for VoIP up to 50% [18].

One other consideration to keep in mind is that, even when UDP or RTP do not retransmit the packets, the HSPA protocol layers will provide sufficient protection and error recovery mechanisms through the MAC layer HARQ functionality. It could also be possible to introduce further protection using the RLC acknowledged mode, however this is normally discouraged because of the little extra gain over the MAC retransmission scheme, and the extra delay added to the frames.

In addition to the protocol type, QoS treatment is very important in ensuring a quality delivery of VoIP service, especially in wireless environments where congestion is likely to happen. In order to ensure an efficient handling of VoIP traffic, the packetized voice traffic is classified as a Real Time Conversational service, which as covered in Chapter 3 (QoS), has stringent delay requirements and somewhat loose data protection requirement. This topic is discussed further in Section 8.4.4.2.

8.4.2.3 Session Control Protocol

A Session Control Protocol is required in order to properly establish the voice calls between two VoIP users, maintain, and terminate them when needed. Such a protocol could be very basic (with only START/STOP commands), or more complex in which case they can deliver advanced telephony services such as call forwarding, call waiting, providing interaction with other existing voice services, etc.

The most used VoIP control protocols are H.323 and SIP. The H.323 standard, introduced by ITU-T in 1996, was the first standard developed to deliver the VoIP technology. In the same timeframe, IEEE introduced SIP as an alternative to H.323, but it didn't get support until a few years later when it was adopted as a 3GPP protocol. In 2001, H.323 was the most widely deployed standard with SIP rapidly gaining ground, and since then most of the IP vendors are supporting SIP over the H.323 standard [20].

We will focus on SIP, because it's the protocol adopted by the 3GPP for its IMS solution. The services offered by SIP include:

- user location;
- user capabilities;

- user availability;
- call set-up;
- call handling;
- call forwarding;
- callee and calling 'number' delivery;
- portable number/address;
- terminal-type negotiation and selection;
- terminal capability negotiation;
- caller and callee authentication;
- blind and supervised call transfer;
- invitations to multicast conferences.

SIP has some limitations regarding the support of some operator's regulatory requirements such as lawful interceptions or emergency calls; however there are solutions available for these requirements and ongoing standardization work that will ensure proper support of these features.

8.4.3 Requirements for VoIP as a Complete Voice Service

This section analyzes the service requirements that VoIP should meet in order to be successfully offered as a commercial service in any network. Throughout this section we identify a 'primary voice service' as a voice telephony service over IP with the same quality, robustness and feature set as current cellular operators are delivering today in circuit-switched mode. Following the same argument, a 'secondary voice service' is that in which a voice communication is possible between two peers with a customer perception of a lower quality service than the one currently received with conventional telephony.

8.4.3.1 Reception Quality

The quality of the contents delivered by the VoIP primary service should be equivalent to existing cellular voice standards, which typically require a MOS of 3.0 or higher. A secondary voice service could afford worse voice quality, in the range of 2.5 MOS. Furthermore, a primary voice service must include the possibility to transport non-voice content (music and DTMF tones) without significant distortion.

The quality of the VoIP service is primarily affected by three factors: the selected vocoder, the network packet loss and the presence of echo:

Vocoder: Since VoIP can use any type of existing vocoder, it can potentially deliver the same quality as GSM/WCDMA if it employs the AMR codec set, or similarly, if the CDMA2000 family of codecs are used. Furthermore, there are other alternatives that can deliver similar voice quality than the two previously mentioned, such as G.723.1 or iLBC.

Packet Loss: Caused by radio link bit errors, network congestion, packet buffer discards, or processor overload in the network. Packet errors are often bursty in nature. Acceptable packet loss limits depends on the vocoder used, and range between 2 and 15%.

Echo: Caused by voice signals being reflected in the network, typically at interface points (e.g. 2- to 4-wire hybrid in analog network, acoustical speaker-microphone on terminal device, etc.). The user does not perceive echo that is sufficiently attenuated or delayed by less than about 15 msec. Echo between 15 and 35 msec results in a hollow sound to the voice audio, however echoes beyond 35 msec are audible and must be cancelled for good quality voice telephony [21]. The echo effect can be, by the additional delays of VoIP, typically in the range of 50 to 100 msec. Good echo cancellation signal processing equipment is required, particularly in the VoIP gateways to the public telephone network.

8.4.3.2 Delay

The overall delay is an addition of several components, including handset and network processing delays, vocoders, radio transmission, transport, etc. Delay can be mitigated by employing QoS in the network design to prioritize voice packets to minimize switching and routing delays. Appropriate packet size selection can lower delays as well.

The ITU-T[2] study on [22] analyzes the effect of the end-to-end delay on user's perception of the communication quality. A very high delay can be quite annoying in a conversation independent of the quality of the waveform being received. According to this study, the ear-to-mouth delay should be below 250 ms in order to be considered a primary service, and up to 300–350 ms for secondary voice services.

In addition to the end-to-end delay, voice quality can suffer due to delay variations of the transmitted packets (jitter). Jitter is managed by jitter buffers in the packet receive path to remove jitter before the voice samples are converted to audio. Jitter buffers may discard packets exceeding a threshold delay, resulting in additional packet loss.

8.4.3.3 Call Setup Time

In order to be an acceptable primary service, VoIP call setup times should be in the order of the typical standards in cellular networks, which range between 4 and 10 seconds, depending on the network setup and destination number. A service with call setup longer than 10 s would be considered a secondary voice service.

8.4.3.4 Mobility and Interworking

A primary voice service should offer the same level of mobility as the existing cellular service, therefore the handover interruption times should be inaudible to the customer (< 300 ms). In addition, it should be possible to keep the voice call in areas where only CS voice service is available, therefore proper handover and interworking mechanisms should be in place in order to avoid a call drop.

---

[2] The **International Telecommunication Union** is an international organization established to standardize and regulate international radio and telecommunications.

Table 8.6   VoIP primary service requirements

Quality (MOS)	E2E delay	Call setup time	Handover interruption	Additional features
> 3.0	< 250 ms	< 10 s	< 300 ms	Current feature set, TTY and TTD, lawful interception and Emergency Calls

8.4.3.5 Calling Feature Set

A primary service should provide an equivalent calling feature set as the one currently offered by cellular networks, including the following: caller ID, call waiting, call forwarding, 3-way calls, etc. Furthermore, a primary voice service needs to support Text Telephony (TTY) and Telecommunications Device for the Deaf (TTD), in addition to special regulatory requirements such as lawful interception and Emergency Calls.

Table 8.6 summarizes the basic requirements that should be satisfied by a VoIP primary service being offered commercially.

8.4.4 HSPA Enablers for Voice Over IP

While VoIP as a service could be offered in today's HSPA commercial networks, it would still not be a competitive service as compared to the current cellular voice in terms of quality, capacity and reliability. Support of VoIP in current HSPA networks is very limited due to performance challenges and inefficiencies in the resource utilization for low bit rate applications.

The introduction of HSDPA/HSUPA is the first step into the support of higher efficiency in the utilization of the resources (both for power and codes), and represents a significant improvement in the user performance, with increased bitrates, and a reduced latency of 70 ms versus approximately 180 ms in Rel.'99. HSPA, however, is not by itself a VoIP efficient technology. There are several challenges to be addressed that affect different aspects such as mobility, battery life, end-user performance and maximum number of users that can be served by one cell. This section discusses the features that will make VoIP a viable service from both performance and capacity points of view.

In this section, new features are discussed that enable VoIP technology to be a serious candidate to substitute for the existing CS voice service. These features are provided in Rel.'7 and Rel.'8. The four major areas for comparison with the current AMR voice service are summarized in Table 8.7.

The following are the major features that enable the practical realization of VoIP in HSPA networks:

- Fractional DPCH (presented in Section 8.2.4.1);
- Uplink Gating (presented in Section 8.2.4.2);

Table 8.7 Comparison between CS Voice and VoIP

Spectral efficiency	Voice quality	Coverage	Mobility
VoIP spectral efficiency will be equivalent or better than Rel'99 AMR with VoIP special improvements included in Rel'7	VoIP quality can be comparable to CS voice as long as it's provided over controlled devices with proper QoS treatment and sufficient processing power	No special concerns on coverage, VoIP has a similar link budget as CS voice	Mobility with CS Voice is superior as with VoIP however VoIP can close the gap through the Voice Call Continuity feature (Rel'8)

- HS-SCCH-less operation (presented in Section 8.2.4.5);
- Robust Header Compression;
- Support of End-to-end QoS;
- Voice Call Continuity;
- Delay Sensitive Scheduler.

The following sections provide details on the new features required for VoIP that were not previously discussed.

8.4.4.1 Robust Header Compression (ROHC)

Header compression techniques have been used for a long time in the TCP/IP world [23,24]; however, ROHC [25] is a new method specifically designed for wireless links, where the packet loss is high. This compression technique can handle several consecutive packet losses while maintaining the same compression efficiency.

The header overhead in typical streaming applications is in the order of 40 to 60 bytes per IP packet, which corresponds to around 60% of the total amount of data in the case of VoIP applications. With ROHC this overhead can be compressed to typically 1 or 3 bytes. If this technique is properly applied, it can effectively double the VoIP capacity, since VoIP packets are typically of very reduced size.

Frame aggregation techniques can also improve the overhead efficiency by grouping frames that are directed to the same user in order to avoid padding during the construction of the radio block [19]. The drawbacks are an increased buffer delay and a potential higher error rate, since several frames will be lost at the same time.

8.4.4.2 Support of End-to-End Quality of Service

Networks with QoS support can handle more VoIP calls with tighter jitter requirements than those without QoS. Current 3GPP networks have in place many of the functions required to implement efficient QoS management, such as admission and congestion control, efficient schedulers, etc. However, there are aspects of the QoS management that have to be specifically configured for efficient support of VoIP services, and in some cases, the existing functionality will need to be enhanced.

In order to achieve a VoIP service quality that could be comparable to the current CS voice, the first requirement from a QoS point of view is the support of conversational PS bearers. The need for a conversational bearer is justified by the tight delay and jitter requirements from the voice service. The VoIP packets cannot afford to be waiting for a long time in the system and therefore the resource management mechanisms in the network have to be aware of this and assign them a high priority in the different scheduling queues. One very important piece in the data path is the transmission over the air interface, therefore an efficient delay-sensitive packet scheduler is recommended as discussed later in this section.

One important aspect to consider, in addition to the packet prioritization, is the capability of stealing resources from other services (preemption). The VoIP service should be configured to allow preemption when congestion occurs either in the air interface, or in any of the interfaces, to avoid blocking of voice calls due to the usage of other less critical data services.

Finally, the QoS scheme needs to be able to negotiate all the network resources along the communication path, this is, to have an end-to-end control. With 3GPP Rel.'7 QoS architecture and its enhancements in Rel.'8, it is possible to deploy functionality to assure the end-to-end delay and quality of the connection.

It is worth noting that sophisticated QoS management will allow the operator to establish trade-offs between the quality experienced by the user and the overall network capacity. For this reason, in order to achieve optimal performance in the network it's important to obtain a good understanding of the minimum performance requirements (BLER, delay, jitter) that should be allowed for in the VoIP service. As an example, in Figure 8.25, extracted from a study on VoIP over HSDPA, illustrates how with loose E2E delay requirements the network capacity can increase up to 40%, from 72 users with 80 ms up to 104 users with 150 ms.

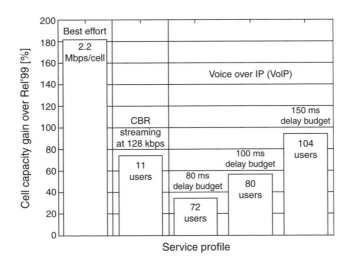

Figure 8.25   Tradeoff between delay and capacity in a VoIP HSDPA network [19] © 2006 IEEE

8.4.4.3 Voice Call Continuity (VCC)

One critical aspect in the deployment of VoIP service is the handover to other networks, or to other bearers within the same network; for instance, transitioning from VoIP/HSPA to CS/Rel'99 in areas where the VoIP service is not available. This functionality, called Voice Call Continuity or VCC, has been discussed within 3GPP for several years and is now being specified as part of Rel'8. VCC will permit:

■ voice calls to be routed over Cellular CS or VoIP bearer based on available radio bearers and subscriber preferences, and
■ handovers between a 3GPP CS system and IMS/VoIP, and vice versa.

8.4.4.4 Delay-Sensitive Scheduler

All the features discussed in the previous subsections represent improvements being introduced in the 3GPP standards to boost the network capacity for PS services. As good as these features are, there needs to be appropriate vendor implementation of the Radio Resource Management (RRM) algorithms to account for special services like VoIP in order to fully exploit all these gains. In particular, it is essential that the scheduler is able to prioritize conversational services based on the delay budget and required jitter.

On the other hand, when a packet needs to be retransmitted several times there is a point after which it's pointless to resend the packet because it will be already beyond the limits of acceptable latency (i.e. it will be discarded as being out of the acceptable delay window for VoIP). In these cases there is a benefit in implementing special mechanisms to discard packets that have reached a certain 'life time'. Packet discarding can be implemented both in the Node-B and other intermediate nodes, and will help reduce delay on the rest of the packets during congestion situations.

Figure 8.26, extracted from a study of VoIP over HSPA Rel.'7 shows the impact of the packet scheduler and 3GPP air interface features on VoIP capacity. The simulations show that up to 60% additional capacity can be obtained through a VoIP aware scheduler compared to the basic Round Robin one. When combined with other enhancements, such as F-DPCH and improved receivers, the capacity gains can be on the order twice or more.

8.4.5 Performance of VoIP in HSPA Networks

This section provides measurements from a VoIP lab test with commercial HSDPA Rel.'5 equipment. At the time of the trials most of the VoIP specific functionalities were not supported by the network; however these results help understand the potential of the technology and the reliability of existing simulation results. At the end of the section we present the capacity and performance expectations from a VoIP/HSPA network with all the bells and whistles as defined in the standards up to Rel.'7.

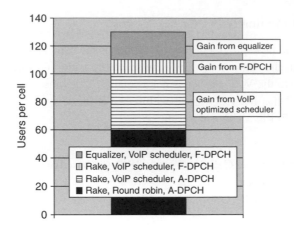

Figure 8.26   Comparison of VoIP capacity for different Schedulers and receivers [26] © 2006 IEEE

8.4.5.1 Preliminary Assessment of VoIP Performance Through Lab Tests

An analysis of the VoIP technology was conducted in a lab under a controlled environment with the objective of gaining knowledge on the *quality* of the voice service being transferred through the packet network, with different codecs and under different conditions. The tests were performed with laptops using HSDPA data cards and a PC VoIP client.

The baseline performance was established through a set of tests with a direct Ethernet connection. With this setting we performed additional test cases using a network impairment tool to simulate the effects of packet loss, jitter and end-to-end delay.

The voice quality was measured with a commercial MOS tool [27]. This tool sends standard voice from one PC to the other. The quality score is derived using the PESQ algorithm [28], which compares the received sample with the original waveform and finds distortion, noise, etc. based on waveform analysis.

We performed a first codec analysis with the direct PC to PC connection, where we found that the quality offered by the VoIP client was similar, if not better, than the one offered by cellular phones. The resource consumption was high, as expected, due to the overhead introduced by IP and Ethernet. Of the analyzed codecs the best tradeoff was achieved with the iLBC codec, as depicted in Figure 8.27 comparing voice quality versus bitrate.

The iLBC codec also showed good performance under severe packet loss, with an average MOS of 3.0 (2.5 MOS at the 95%), as depicted in Figure 8.28. The simulated packet loss was bursty rather than uniform in order to emulate congestion situations in which several consecutive packets may be lost. The codec response to jitter was also assessed with jitter samples following an exponential distribution, although no significant difference was found between the GSM and the iLBC codec.

After these preliminary tests we used the iLBC codec for the remainder of our lab tests because it was the one showing the best overall performance with VoIP. The second phase of the

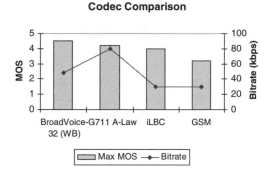

Figure 8.27   Comparison of voice quality offered by different vocoders with VoIP © 2006 IEEE

lab test was conducted with commercial HSDPA equipment inside a lab with a controlled radio environment under which serving signal strength, fading and interference were fully controllable. Our setup also allowed us to perform mobility tests between two different HSDPA cells. Due to limitations on our fading simulator we could only simulate fading on the downlink, as shown in Figure 8.29.

The VoIP communication was established between two laptops that were attached to the UMTS/HSPA network. A MOS tool provided real-time scoring of the samples. At the point of the test the HSUPA technology was not available; therefore the UL channel was using Rel.'99 radio bearers. For each scored sample, the voice was transmitted from one PC using UMTS Rel.'99 and was received at the other PC through HSDPA. See Figure 8.30 for a diagram of the lab setup.

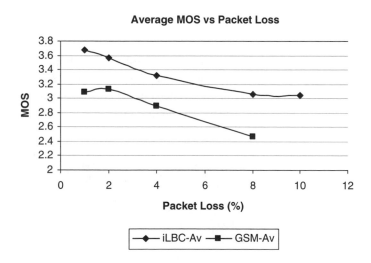

Figure 8.28   Codec comparison under packet loss conditions (iLBC vs. GSM-FR) © 2006 IEEE

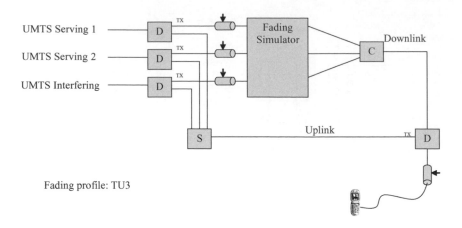

Figure 8.29   Diagram of radio environment setup

The network was based on 3GPP Rel.'5 and its capabilities were very limited with regards to the required radio features for an optimum PS voice performance. The test setup had no support of guaranteed bitrates, QoS scheduler, header compression or any of the Continuous Packet Connectivity features. Furthermore, HSDPA mobility was not working properly at the time of the tests. In summary, the results need to be interpreted as a very preliminary assessment of the VoIP quality in a HSDPA network.

With such a test setup, we evaluated the performance of VoIP under different signal strength and fading conditions. The signal at the antenna was attenuated gradually from -70 dBm up to -110 dBm, with TU3, TU50 and Pedestrian A fading models. The results, shown in Figure 8.31, indicated good voice quality (MOS > 3.0) in all the cases until the packet session could not be sustained anymore, which occurs around RSCP $= -110$ dBm. The measured end-to-end delay was higher than expected at around 300 ms. An analysis of the transmission on the RAN indicated that most of this delay was caused in the PCs, probably by the VoIP software or the laptop processor itself.

Figure 8.31 also illustrates the effect that uncontrolled data traffic has on the VoIP performance. The last line of the chart (Mux) models the case where the two VoIP users share the HSDPA bandwidth with an FTP user. Since this network did not implement any QoS support, when the radio conditions are degraded and the HSDPA bitrate gets stressed, the VoIP

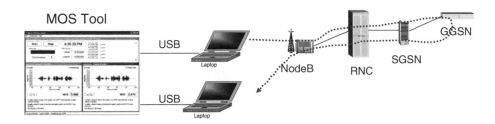

Figure 8.30   VoIP lab setup

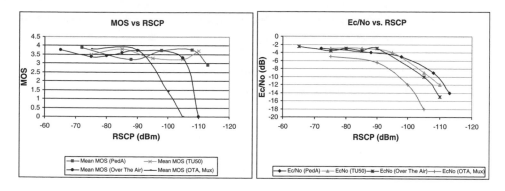

Figure 8.31 MOS results with different signal strength (left) and corresponding Ec/No (right)

users don't get preferential treatment over the FTP download. The VoIP users' performance becomes poor when RSCP level reaches -100dBm, 10 dB before the unloaded case.

Finally, we performed some initial assessment on VoIP quality under mobility conditions. It was expected that HS-DSCH cell changes occurring inside the same RNC would have a very low interruption time, in the order of 200 ms or lower. We observed unexpected behavior as the HSDPA users approached a cell transition: the MOS value was noticeably reduced every time that there was a Radio Link Addition or Deletion procedure, as shown in Figure 8.32.

In summary, the initial assessment after lab tests indicated that VoIP can deliver good voice quality in a HSDPA environment if certain features are implemented to improve mobility performance and reduce end-to-end delay. As noted earlier, all of these issues are properly addressed with the Rel'7 and Rel'8 standard features.

8.4.5.2 Expected VoIP Capacity Based on Simulation Results

After having reviewed all the enhancements provided by Rel.'7 and Rel.'8, including spectral efficiency, latency and mobility improvements, and having experienced the quality of the voice,

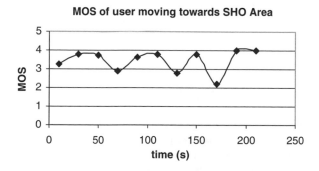

Figure 8.32 VoIP performance in the Soft-Handover areas

we believe that the VoIP technology will become a solid and pervasive service in HSPA networks in the medium term (approximately year 2010). At that point, with all necessary features in place, it is expected that the voice capacity of the sectors will increase by 100% compared to Rel.'99 channels, from 60 to 120 users/cell. At this capacity limit, actually the bottleneck could become the hardware equipment in the Node-Bs.

8.5 Summary

Several UMTS service providers pushed to create an HSPA evolution called HSPA+ (or HSPA evolution) with the objective to build an intermediate stepping stone that brings some of the benefits of 4G but based on an HSPA foundation. HSPA+ was conceived to reduce latencies, increase peak rates, reduce battery consumption and increase capacities. Smooth inter-working from one technology to the next, backward compatibility and an easy upgrade path were practical considerations too. The main goals and proposed solutions for HSPA evolution were as follows:

- A flat architecture and signaling enhancements designed to reduce latency.
- Conceptually peak data rates can be doubled on the downlink through multiple antenna MIMO technology.
- Discontinuous receive (DRX) combined with gating (slotted mode) can reduce battery consumption.
- Advanced receiver processing, including equalization, mobile diversity, interference cancellation, and Higher order modulation (64QAM, 16QAM) can be used to increase cell throughputs and capacities.
- Continuous packet switched connectivity supports a large number of 'always on' packet data users. This prepares the HSPA+ network for applications like widespread use of VoIP, where multitudes of handsets would need to be simultaneously connected to the network.

Features supported in HSPA+ are expected to reach deployment status in commercial networks in the 2009–2010 timeframe. Many of these features come in the form of software upgrades, although some of them (such as MIMO) will require hardware upgrades. The network enhancing features that require hardware upgrades can be added on an as-needed basis, such as in hot spots, rather than across all of the network.

References

[1] 3GPP. RAN Technical. Report R2-060493, 'Scope of Future HSPA Evolution', 3G Americas, February 2006.

[2] 3G Americas, 'UMTS Evolution from 3GPP Release 7 to Release 8. HSPA and SAE/LTE', July 2007.

[3] 3GPP Technical Report 25.876, 'Multiple Input Multiple Output (MIMO) antennae in UTRA'.

[4] 3GPP Technical Specification 25.101, 'User Equipment (UE) radio transmission and reception (FDD)'.

[5] Nihtila, T., Kurjenniemi, J., Lampinen, M., Ristaniemi, T.;'WCDMA HSDPA network performance with receive diversity and LMMSE chip equalization', Personal, Indoor and Mobile Radio Communications, 2005. PIMRC 2005. IEEE 16th International Symposium Volume 2, 11–14 Sept. 2005 Page(s):1245–1249 Vol. 2.

[6] 3GPP Technical Report R4-061068, 'Some observations on DIP values as a function of network geometry'.

[7] 3GPP Technical Report R4-060958, 'Comments on the Interference Cancellation (IC) study item simulation'.

[8] Wanstedt, S., Ericson, M., Hevizi, L., Pettersson, J., Barta, J.; The effect of F-DPCH on VoIP over HSDPA Capacity; Vehicular Technology Conference, 2006. VTC 2006-Spring. IEEE 63rd, Volume 1, 2006 Page(s): 410–414.

[9] 3GPP Technical Specification 25.903, 'Continuous connectivity for packet data users'.

[10] 3GPP Technical Report 23.809, 'One Tunnel solution for Optimisation of Packet Data Traffic'.

[11] 3GPP Technical Specification 23.060, 'General Packet Radio Service (GPRS); Service description; Stage 2'.

[12] 3GPP Technical Specification 29.060, 'General Packet Radio Service (GPRS); GPRS Tunnelling Protocol (GTP) across the Gn and Gp interface'.

[13] 3GPP Technical Report 25.999, 'High Speed Packet Access (HSPA) evolution; Frequency Division Duplex (FDD)'.

[14] 3GPP Technical Specification 29.212, 'Policy and charging control over Gx reference point'.

[15] 3GPP Technical Report 23.802, 'Architectural enhancements for end-to-end Quality of Service (QoS)'.

[16] Whitepaper 'iLBC – designed for the future', Global IP Solutions.

[17] Whitepaper 'Voice quality over IP based networks', Electronic Communications Committee of CEPT, Gothenburg 2004.

[18] Rittenhouse, G., Haitao Zheng; Providing VOIP service in UMTS-HSDPA with frame aggregation; Acoustics, Speech, and Signal Processing, 2005. Proceedings. (ICASSP '05). IEEE International Conference on, Volume 2, 18–23 March 2005 Page(s): ii/1157–ii/1160 Vol. 2.

[19] Pedersen, K.I., Mogensen, P.E., Kolding, T.E.; QoS Considerations for HSDPA and Performance Results for Different Services; Vehicular Technology Conference, 2006. VTC-2006 Fall. 2006 IEEE 64th, Sept. 2006 Page(s): 1–5.

[20] Whitepaper 'An Overview of H.323-SIP Interworking', Radvision 2001.

[21] Chan T.Y., Greenstreet D. et al. 'Building Residential VoIP Gateways: A Tutorial'.

[22] ITU-T Recommendation G.114, 'One-way transmission time'.

[23] IETF RFC 1144, 'Compressing TCP/IP Headers for Low-Speed Serial Links'.

[24] IETF RFC 2508, 'Compressing IP/UDP/RTP Headers for Low-Speed Serial Links'.

[25] IETF RFC 3095, 'RObust Header Compression (ROHC): Framework and four profiles: RTP, UDP, ESP, and uncompressed'.

[26] Harri Holma, Holma, H., Kuusela, M., Malkamaki, E., Ranta-aho, K.; VOIP over HSPA with 3GPP Release 7; Personal, Indoor and Mobile Radio Communications, 2006 IEEE 17th International Symposium on, Sept. 2006 Page(s): 1–5.

[27] http://www.metricowireless.com/.

[28] ITU-T Recommendation P.862, 'Perceptual Evaluation of Speech. Quality', ITU-T, February 2001.

9

Technology Strategy Beyond HSPA

In the last few years a new 3GPP technology called 'Evolved UTRAN', also known as UMTS 'Long Term Evolution' (LTE) or Evolved Packet System (EPS) has been standardized, which puts the UMTS operators in the position to decide how they will migrate their current networks: should they join the UMTS 'evolutionary' path (HSPA+), the 'revolutionary' path (LTE), or both?

LTE technology introduces many performance benefits for the operator: flexible spectrum bandwidth, higher peak data rates and lower latency. LTE has been designed to be a more efficient, scalable network with a simpler architecture which will result in reduced costs (CapEx and OpEx) in the long term for the operators. Section 9.1 provides a basic overview of this new technology.

As was demonstrated in Chapter 8, HSPA technology has been evolving too, to the point that it provides a viable and compelling alternative to LTE in the next five-year timeframe. Section 9.2 compares the expected performance of LTE against that of HSPA and HSPA+, with interesting findings regarding cell capacity and typical user throughputs.

In the long run, there is no doubt that the LTE technology will be the ultimate migration path for wireless operators. There are multiple reasons for this, but one of the strongest one is the operators' desire to make this technology a success. All the major 3GPP standards development efforts are concentrated on developing LTE, and surprisingly major 3GPP2 operators such as Verizon have decided to join LTE as well. The technology will be further improved with the introduction of LTE Advanced, the evolution of LTE that is currently under discussion. The ecosystem will be a major driver, too, enabling global roaming and reducing the price of the equipment due to the economies of scale.

However, what is unclear is when it makes business sense for an operator to switch to LTE. Although much effort has been put into LTE development, technologies take time to be standardized, implemented into reliable, commercial equipment and more importantly, reach a maturity where a healthy portfolio of terminal devices can be offered to the consumer. Section 9.3 reviews the technology timelines considering all these aspects, and as we'll see

HSPA Performance and Evolution   Pablo Tapia, Jun Liu, Yasmin Karimli and Martin J. Feuerstein
© 2009 John Wiley & Sons Ltd.

the capabilities of LTE will be limited to PC card broadband offering during the first few years after launch. In the meantime the cellular industry has seriously shifted from voice towards mobile data, and operators need to have a tool to capture that proven and growing revenue stream. The HSPA+ technology can provide the necessary broadband capacity and quality for years to come, allowing the operator to gain a better return on investment before deploying a completely new network. It also provides a smooth migration path to LTE allowing a stepped transition. Section 9.3 reviews different evolution paths for operators, analyzing the pros and cons of each of the alternatives.

9.1 Introduction to Evolved UTRAN

In November 2004 several manufacturers, operators and research institutes presented contributions to the 3GPP RAN Evolution workshop in Toronto, recommending a new 3GPP technology that could provide a solid alternative to other emerging technologies, such as WiMAX. At that point, WiMAX technology was perceived as a significant threat to UMTS, promising a wireless broadband experience with a flat, simplified network architecture and reduced cost per bit. Considering the data consumption trends in the broadband industry, with a typical residential internet customer consuming an average of 2–3 GB/month of data, a much cheaper network infrastructure was needed for 3GPP operators to be able to compete in the fierce broadband environment.

At that time, the technical specifications of the HSDPA technology were finalized, but the first commercial network wouldn't become a reality until one year later, and the 3GPP members didn't believe that HSDPA would be sufficient to satisfy their long term needs of higher bitrates with a lower cost. In 2004, the UMTS technology had failed to deliver a sufficiently compelling broadband experience, with data bitrates and latency comparable to what an EDGE network could provide. Furthermore, the 3G penetration was low even after three years of having deployed the networks, because of the lack of breadth in terminal products and their premium price as compared to 2G handsets. One of the reasons for the higher terminal prices was blamed on the high intellectual property rights costs incurred by handset manufacturers, which in some cases could represent up to 6% of the cost of the terminal [1]. However the main reason was the immaturity of the WCDMA technology itself, which took several more years to materialize.

The development of LTE had the goal of mitigating all the problems encountered in UMTS technology, and was specifically designed to identify a network technology that would meet several stringent criteria, including reduced cost per bit, deployment flexibility and better service performance. In order to achieve these high-level criteria a set of specific performance goals were established. The air interface technology and features would be selected based on the ability to achieve these targets with the lowest complexity and cost [2]:

- Better service performance:
    o significantly increased peak bit rates: up to 100 Mbps in DL, and up to 50 Mbps in UL;
    o increased average throughput: 3 to 4 times Rel.'6 HSPA in DL, 2 to 3 times in UL;

o increased edge of cell bitrates: 2 to 3 times Rel.'6 HSPA;

o reduced latency in the user plane of up to 10 ms.

■ Reduced cost per bit:

o increased DL spectral efficiency by 3 to 4 times, compared to HSDPA Rel.'6;

o increased UL spectral efficiency by 2 to 3 times, compared to HSUPA Rel.'6.

■ Spectrum flexibility, easy to deploy in new and existing bands:

o support for bandwidth allocations of 1.25, 2.5, 5, 10, 15 and 20 MHz.

■ Simplified architecture for OpEx and CapEx savings.

■ Attractive terminal complexity, cost and power consumption.

In parallel with the LTE standardization work, an operators' forum called 'Next Generation Mobile Networks' (NGMN)[1] was created to exercise tighter control over the new network technology that was being defined because the standardization forums typically were controlled by the manufacturers. The NGMN forum created a set of requirements for the next generation network, and included three candidate technologies: LTE, WiMAX and UMB. The NGMN requirements have been considered during the LTE specification work; in particular new requirements were defined to simplify the network deployment and configuration: Self-Organization, Self-Configuration and Self-Optimization.

The LTE standardization work started in 2005 and is scheduled to be completely finalized by 2009. The standardization was divided in three different parts: (1) Radio Access Evolution, called LTE or *E-UTRAN* (Evolved UMTS Radio Access Network), (2) Core Network Architecture Evolution, called SAE or *EPC* (Evolved Packet Core) and (3) Conformance test specifications. The Core Specifications, including Radio Access and Core Network architecture standards are expected to be finalized by the first half of 2009. The conformance specifications, needed for commercial handset development, are targeted to be finalized during 2009, although this timeline could well slip into 2010. Figure 9.1 illustrates the proposed LTE commercialization timelines.

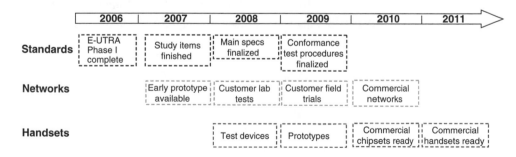

Figure 9.1 Overview of LTE technology timelines

---

[1] http://www.ngmn.org/.

9.1.1 Technology Choice and Key Features

The feasibility study of LTE [3] concluded that the targeted performance goals would require the utilization of a new air interface technology, OFDM, with an All-IP, flat network architecture. The Circuit Switch core was not considered necessary since the voice service could be offered through Voice over IP (VoIP) technology in a PS-only network.

The following are the main characteristics of the new LTE system:

- Air interface technology based on OFDMA and SC-FDMA for downlink and uplink, respectively.
- PS-Only core network (voice support through VoIP).
- No soft-handover: hard-handover with downlink data forwarding on SDU level.
- Simplified E-UTRAN architecture:
  - only one type of node: Evolved Node-B or eNB;
  - eNB controls RRC, scheduler, HARQ, outer ARQ and IP header compression;
  - RLC and MAC layers also terminated in eNB.

9.1.1.1 Main LTE Nodes

LTE significantly simplifies and flattens the current architecture of the access network by reducing the numbers of elements and interfaces. This is possible due to the elimination of the CS side, together with a more optimized PS network layout that decentralizes the control of the network by putting more functionality in the lowest level node of the access network (eNode-B), as shown in Figure 9.2. Such an approach is in line with the promised simplified architecture, reduced latency and reduced cost per bit targets.

The eNode-B is the element controlling all the aspects related to management of radio resources and user plane mobility, while the Access Gateway (aGW) includes functionality for control of mobility management, idle mode, inter-system mobility and legal interception.

The eNode-Bs are interconnected through the X2 interface. The X2 interface allows the eNode-Bs to autonomously handle handovers, without oversight from a higher layer element

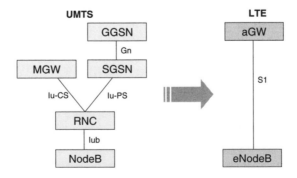

Figure 9.2   Radio Access architecture evolution

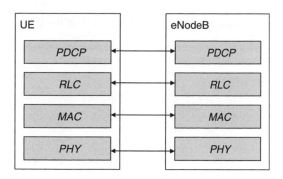

Figure 9.3   LTE User Plane protocol architecture

such as RNC or BSC. The X2 interconnects are an important aspect allowing the flatter network architecture with more functionality embedded in the eNode-B.

9.1.1.2 Protocol Architecture

Figure 9.3 illustrates the protocol architecture of the user plane of E-UTRA. The functionality of each of the layers is similar to that of the HSPA architecture, except for two major changes:

1. Both the MAC and RLC layers are terminated in the eNode-B, therefore the outer ARQ mechanisms, admission and load control are located in the eNode-B.
2. The PDCP layer, which includes functions like IP header compression, is terminated in the eNode-B.

With regard to the control plane the main change compared to the HSPA architecture is the termination of the RRC protocol in the eNode-B, as shown in Figure 9.4. The RRC performs broadcast, paging, RRC connection management, radio bearer control, mobility functions, UE

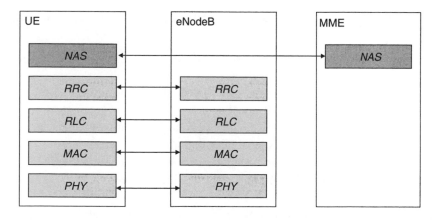

Figure 9.4   LTE Control Plane protocol architecture

measurement reporting and control. Note that the control of the connection mobility is now fully done by the eNode-B.

9.1.1.3 Key Radio Features

The key features that have been included as part of the E-UTRAN specifications are:

- *OFDM technology in downlink.* OFDM provides improved channel orthogonality compared to code division multiple access schemes, which results in an overall capacity gain ranging between 20% and 70% over WCDMA.
- *SC-FDMA technology in uplink.* SC-FDMA technology permits the orthogonality gains from OFDM with a lower Peak to Average Power Ratio (PAPR). A lower PAPR results in better power consumption in the terminal and likely on a lower cost of the devices.
- *Support of MIMO.* MIMO significantly increases the user bitrates and can provide a reasonable capacity gain (in the order of 20%). Different MIMO schemes are supported in uplink and downlink, with configurations of up to four transmit and four receive antennas. Single user and multi-user MIMO are also supported.
- *Frequency domain scheduler.* OFDM users can receive in different sub-channels at different times, thus improving the robustness of the link against fading. Time and frequency domain schedulers can provide a capacity gain of 40% over normal schedulers.
- *Higher Order Modulation.* LTE supports transmissions with up to 64QAM, greatly improving the peak bitrates in favorable radio conditions.
- *Short Time-to-Transmit (TTI) interval.* The LTE frame duration is just 1 ms, which allows a significant reduction in the radio latency.
- *Fast link adaptation and HARQ.* These ensure that the links adapt to changing radio conditions, improving the decoding performance in unfavorable radio conditions.

9.1.2 Architecture and Interfaces

Figure 9.5 illustrates the overall E-UTRAN architecture in the context of other 3GPP technologies [4].

Below is a description of each of the elements and interfaces related to the packet core:

MME – Mobility Management Entity – in charge of authentication and authorization, management of UE context and UE identities.

UPE – User Plane Entity – terminates the data path, triggers/initiates paging, manages and stores UE contexts, and facilitates legal interception.

IASA – Inter Access System Anchor – combines the 3GPP anchor and the SAE anchor.

3GPP Anchor – keep control of the user plane for mobility between the 2G/3G access system and the LTE access system.

SAE Anchor – keep control of the user plane for mobility between 3GPP access systems and non-3GPP access systems.

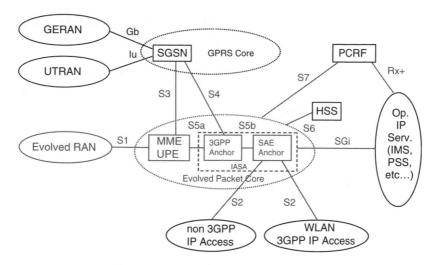

Figure 9.5   E-UTRAN Packet core architecture [4]

S1 – provides transport of user plane and control plane between the eNode-Bs and the Evolved Core Network.

S2 – provides with user plane control and mobility support between IP access networks (such as WLAN) and the SAE Anchor.

S3 – enables the integration of legacy GPRS core, providing an equivalent to the Gn reference point as defined between SGSNs, for exchange of user and bearer information for inter 3GPP access system mobility.

S4 – user plane control and mobility support between GPRS Core and the 3GPP Anchor. It is based on the Gn reference point as defined between SGSN and GGSN.

S5a – user plane control and mobility support between MME/UPE and 3GPP anchor.

S5b – user plane control and mobility support between 3GPP anchor and SAE anchor.

S6 – interface between the EPC and the Home Location Register. Enables the exchange of authentication and authorization information to control user access.

S7 – equivalent to the Gx interface in UMTS. It facilitates the exchange of information between the Policy Control Resource Function (PCRF) and the Policy and Charging Enforcement Point (PCEP).

SGi – reference point between the Inter AS Anchor and the external packet data network. This reference point corresponds to the Gi functionality in GPRS and supports any 3GPP and non-3GPP access systems.

9.1.3 Early LTE Trials

In May 2007 several mobile operators and infrastructure vendors created a joint initiative [5] to provide an early learning experience on the new technology being developed. The objectives of the forum are to demonstrate the capabilities of the LTE technology, promote it and drive its commercialization. Its test networks are also useful to perform early equipment interoperability (IOT) tests.

In February 2008 the forum reported peak data rates of up to 300 Mbps with 4 × 4 MIMO, and up to 100 Mbps with baseline devices. RAN latency measurements were also below the target 10 ms specified in the standards.

The forum's activities will now start a series of interoperability tests and continue with trials until the end of 2009, when commercial LTE networks are expected to be deployed.

9.2 Analysis of HSPA vs. LTE

When the work on LTE started in 2005 the wireless industry took an unprecedented approach to define the new standard: a set of performance goals were defined relative to the existing 3G service at the time (Rel.'6 with Rake receiver). The performance objectives covered different aspects, from spectral efficiency to latency and user data rates. Table 9.1 summarizes the LTE performance targets as specified in [2].

As introduced in the previous section, in order to achieve those ambitious targets the E-UTRA standard was defined based on a new air interface technology (OFDM), a very simplistic, flat, PS-only core network, and a set of enhancement features that would help achieve the required goals. When the standard was largely defined a large number of simulations were performed by several members of 3GPP, comparing UMTS Rel.'6 with LTE under a fixed set of conditions [6,7]. The most typical combinations tested are listed in Table 9.2.

A short time after the LTE study item was introduced, the HSPA evolution (HSPA+) initiative emerged with the intent to close the gap between the expected LTE performance and capacity and that of the 'legacy' HSPA networks, many of which were not even deployed at the time. At the time, it was also highlighted that from an operator's perspective the path to LTE may not be the best option in the short to medium term, considering that:

- the new technology would require new spectrum or carving from existing frequencies;
- the large amount of band plans and technologies to be supported in a single terminal would increase the price of the equipment; and
- the higher peak rates would only be realized with an expensive deployment of transport resources.

Table 9.1 Summary of LTE performance goals

	Downlink	Uplink
Spectral Efficiency	3–4 × Rel'6	2–3 × Rel'6
Peak data rate (20 MHz)	100 Mbps	50 Mpbs
User throughput at cell edge	2–3 × Rel'6	2–3 × Rel'6
Latency due to the radio network	10 ms
Idle to active latency	100 ms
Bandwidth	Flexible 1.25–20 MHz

Table 9.2 Typical test scenarios for LTE performance bechmarking

	Downlink	Uplink
MIMO configuration (LTE only)	2×2 SU-MIMO, 4×2 SU-MIMO, 4×4 SU-MIMO	1×2, 1×4, 1×2 SU-MIMO
Scenario	Macro cell 500 m ISD, 1732 ISD	

On the performance and capacity side, the HSPA+ initiative wanted to establish a set of performance goals to be achieved before any specific feature was discussed. These goals are summarized in Table 9.3 as proposed by Cingular (now AT&T); however these were never made official into the 3GPP standards.

As can be seen from Table 9.3, the HSPA+ initiative had very ambitious goals to improve the spectral efficiency of the network, and was not so much focused on the peak data rates or scalable bandwidth benefits of LTE. The idea was to achieve these goals without a significant impact to the hardware that was already deployed. As a matter of fact, HSPA+ would benefit from many of the concepts introduced by LTE such as MIMO and higher modulation orders (64QAM for DL and 16QAM for UL). Table 9.4 compares the features included in both technologies (up to Rel.'8).

As can be observed from Table 9.4, many of the enhancing features from LTE are also present in the evolution of HSPA. The two major differences are (1) the air interface technology, which is OFDM instead of CDMA, and (2) the simplified network architecture.

The OFDM technology is believed to be more efficient than CDMA due to the improved orthogonality, however, the real gains from the technology come from the Frequency Domain Packet Scheduling, which is not possible to develop in CDMA. On the latency side, LTE is clearly designed to present lower latencies due to the shorter frame length and optimized network architecture; however HSPA+ has made significant improvements compared to Rel.'6 with the introduction of the Evolved HSPA architecture.

Today, with Rel.'7 standardization completed and Rel.'8 items well under way, the peak performance of HSPA+ has significantly improved with respect to the Rel.'6 baseline adopted by 3GPP for the LTE comparison. Table 9.5 compares the peak throughputs and latency achieved by HSPA+ and LTE for configurations with up to 2×2 MIMO. Higher MIMO cases are not considered in the comparison due to the difficulty of having more than two antennas either in a tower or in a mobile device.

Table 9.3 HSPA+ performance objectives proposed by Cingular

	Downlink	Uplink
Spectral Efficiency	$3–4 \times$ Rel'6	$1.2–3 \times$ Rel'6
User throughput at cell edge	$3–4 \times$ Rel'6	$1.2–3 \times$ Rel'6
Latency due to the radio network	40 ms	
Idle to active latency	100 ms	

Table 9.4 Comparison of enhancement features (LTE vs. HSPA+)

	HSPA+	LTE
Air interface	CDMA, TDMA (DL)	OFDM, TDMA (UL + DL)
Flexible bandwidth	DL only: 5, 10 MHz	UL + DL: 1.4, 3, 5, 10, 15, 20
Frame length	2 ms	1 ms
HARQ	Chase & IR	Chase & IR
DL MIMO	2 × 2	2 × 2, 4 × 2, 4 × 4
UL MIMO	n/a	1 × 2 MU-MIMO, 2 × 2 SU-MIMO
CS voice support	yes (CSoHS)	No
VoIP support	yes	Yes
DL IC	yes	Yes
UL IC	yes	Yes
FR Scheduler	No	Yes
Power control	UL yes, not in DL	Yes
UL Modulation	QPSK, 16QAM	QPSK, 16QAM, 64QAM
DL Modulation	QPSK, 16QAM, 64QAM	QPSK, 16QAM, 64QAM
Flat architecture	partial (GPRS One-Tunnel, Evolved HSPA architecture)	Yes

9.2.1 Performance Comparison of LTE vs. HSPA Rel.'6

In terms of overall capacity, it is not easy to establish a comparison between both systems because the numbers can only be provided by network simulations, and the results vary greatly depending on the simulation assumptions. In fact, numerous simulation results have been supplied during the LTE standardization process, comparing HSPA and LTE, which showed that the expected performance gains from LTE were being realized. One key assumption of this simulation work is that the HSPA benchmark network is Rel.'6 with G-Rake terminals. Figure 9.6 is a summary of the results from [8]. From the simulation results for a 500 m inter-site distance scenario the LTE expected performance met the target capacity gains established by the standard (3.2 times DL capacity and 2.3 times UL capacity). Again, only cases with a maximum of two antennas have been considered for practical deployment reasons.

LTE also achieves the promised performance on the user experience front, with both greatly improved average and cell-edge user throughputs, as shown in Figure 9.7.

Table 9.5 Comparison of peak performance (HSPA+ vs. LTE)

	HSPA+	LTE
DL Peak data rate	42 Mbps (5 MHz, 2 × 2)	50 Mbps (5 MHz, 2 × 2)
	42 Mbps (10 MHz, 1 × 1 Dual Carrier–Rel'8)	100 Mbps (10 MHz, 2 × 2)
		180 Mbps (20 MHz, 2 × 2)
UL Peak data rate	11 Mbps (5 MHz)	25 Mbps (5 MHz)
		50 Mbps (10 MHz)
		90 Mbps (20 MHz)
Idle to active Latency	<100 ms	<100 ms
Latency	<30 ms	<10 ms

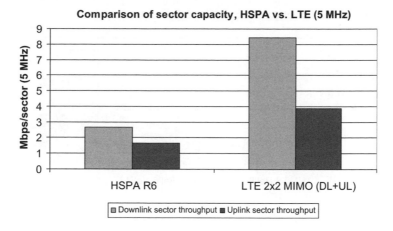

Figure 9.6 Spectral Efficiency comparison between HSPA Rel.'6 and LTE for 500 m ISD (Average of all contributions)

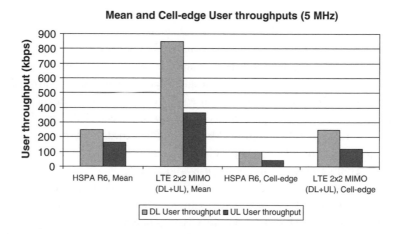

Figure 9.7 Comparison of user experience, HSPA Rel'6 vs. LTE

The last comparison point is the voice capacity supported with an AMR 12.2 kbps codec. The LTE VoIP results vary wildly depending on the vendor, the number of signaling entities (PDCCH), the scheduling strategy (dynamic vs. semi-persistent) and whether packet bundling was used or not. Figure 9.8 compares the voice capacity of UMTS Rel.'6 with two different alternatives, CS voice and VoIP, vs. LTE on a 5 MHz sector. As can be observed, the LTE voice capacity is around three times as high as the limit achieved with either CS or VoIP in Rel.'6.

9.2.2 Performance Comparison of LTE vs. HSPA+

Note that LTE simulation work in 3GPP has always been referred to the performance of Rel.'6 HSPA. Early simulation work [9] showed that the differences in expected performance and

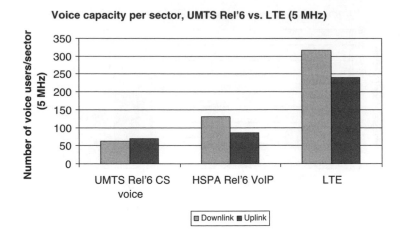

Figure 9.8   Comparison of voice capacity, UMTS Rel.'6 vs. LTE

capacity between HSPA and LTE would be significantly reduced when HSPA deployed the advanced features that have been standardized in Rel.'7 and Rel.'8.

One key aspect of the differences in performance is the fact that the HSPA Rel.'6 results analyzed assumed a Rake receiver instead of an advanced receiver with equalizer. Today, most (if not all) of the HSDPA devices include an equalizer, which is an inexpensive, software upgrade that boosts the receiver performance by between 1 and 4 dB, resulting in overall capacity gains of between 15% and 50%. Also, laptop data cards are typically equipped with two antennas to provide receive diversity. When the HSDPA equalizer is combined with RX diversity, the overall gains are in the order of 80–90% [10].

In the case of voice, the releases prior to Rel.'7 were not designed to carry VoIP traffic efficiently. Rel.'7 introduced a set of features to improve the support of low bitrate applications such as HSPA, including DTX/DRX, HS-SCCH-less operation, UL gating and IP Header compression (ROHC). All these features significantly improve the capability of the HSPA radio network to support VoIP. Furthermore, the Evolved HSPA architecture will help reduce the end-to-end packet latency, which will boost the overall VoIP capacity.

Apart from these HSPA-specific enhancements, as previously mentioned HSPA+ has adopted many of the improvement features contained in LTE, such as MIMO, 64QAM and even the possibility to transmit over 10 MHz of spectrum with the Dual Carrier feature specified in Rel.'8.

With all the different configuration alternatives it is hard to find comparison results that are not biased towards one side or the other, depending on the message the infrastructure vendor or researcher wants to deliver. After having analyzed extensive material provided to us by different infrastructure vendors we believe that the simulations performed by a major 3GPP contributor [11] and shown in Figure 9.9 are a fair estimation of what performance can be expected by HSPA+ compared to LTE, under the same conditions. The HSPA+ simulations

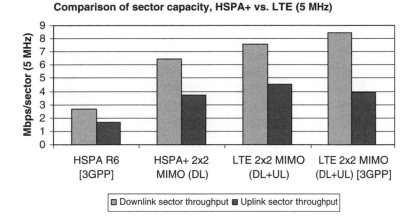

Figure 9.9 Comparison of sector capacity, HSPA+ vs. LTE (5 MHz)

were run under the same scenarios defined for LTE in the 3GPP specifications [6,7]. The
following HSPA+ assumptions were applied:

- scenario: 500 m inter-site distance, mobile speed at 3 km/h;
- 10 simultaneous HSPA users per sector, per 5 MHz (5 DL, 5 UL);
- HSPA+ receivers with either RX diversity + Equalizer or MIMO + Equalizer. No DL
  Interference cancellation assumed;
- HSPA+ performance includes Node-B interference cancellation;
- modulation: up to 64QAM assumed in DL, UL only uses QPSK.

As it can be observed from Figure 9.9, the LTE capacity results from the simulations we
present are comparable with what has been reported by 3GPP, with slightly less downlink
capacity and slightly better uplink capacity. Also, the HSPA+ performance is remarkably close
to that of LTE, which now presents roughly 20% capacity gain compared to HSPA+ both in
uplink and downlink. Furthermore, the LTE uplink simulations assume the utilization of
Interference Management control via the X2 interface, something that most vendors are not
deploying, at least in their initial LTE products. Simulation material from other vendors show
up to 40% spectral efficiency gain from LTE compared to HSPA+, which in any case is far away
from the 200% gain that was claimed when the technology was first introduced.

More surprising are the results of the analysis of the end-user experience with the HSPA+
enhancements. The user throughput at the cell edge (5% of the distribution) is shown to be
similar to that of LTE. In this case the results we present are not in line with the 3GPP
submissions in general, because the number of users per sector is halved; however what's
important is the relative difference between HSPA+ and LTE from the new simulation results in
Figure 9.10. Note that these results also depend on the scheduling strategy defined. The
simulations show that it is possible to achieve a similar user experience in HSPA+ in the less
ideal locations, compared to LTE.

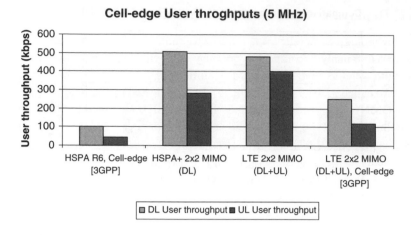

Figure 9.10   Comparison of cell-edge user throughput, HSPA+ vs. LTE

Finally, the voice capacity with HSPA+ is significantly improved with respect to HSPA Rel.'6, due to latency enhancement features that significantly boost the uplink capacity. Figure 9.11 incorporates results based on 3GPP simulation assumptions (3GPP refers to the consensus version agreed during the LTE standardization). Results from a separate study [12] are also included to provide a comparison point for a partial HSPA+ diployment without MIMO or HOM but note that these were obtained under different simulation assumptions.

In absolute terms, it is expected that HSPA+ VoIP capacity practically will double that of the existing UMTS circuit switched voice, for the AMR 12.2 kbps service. In terms of relative capacity with respects to LTE there are significant discrepancies, which range from 10% to 100% capacity gain. All in all, HSPA is not expected to support more than 128 simultaneous users per sector due to hardware capacity limitations on the network side.

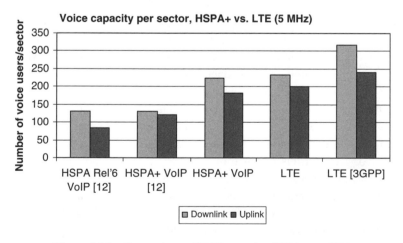

Figure 9.11   Comparison of VoIP capacity, HSPA+ vs. LTE

9.3 LTE Deployment and Migration Scenarios

In today's technology world, the cycle for new technology development is becoming shorter and shorter. While many operators are still busy deploying their 3G networks, the industry, including both vendors and operators, has shifted its focus towards the next generation technology (so called 3.9G or 4G technology). Being selected by the Next Generation Mobile Network (NGMN) as one of its 4G solutions, the LTE/SAE standard is being finalized and trials have started in early 2008. It is expected that the first commercial LTE/SAE network will be available as early as 2010, although a wide availability of terminal devices is not expected until a couple of years later (if not more).

The network technology choice in the case of green field operators is more or less clear – LTE – since the time required to deploy a whole new network will be in line with the maturity of the new technology. However, this is not the most typical case because there are not many green field deployment scenarios left around the globe. The focus of this chapter is on those wireless carriers with an existing network that are analyzing the deployment of LTE. This represents the most generic case, and the most complex as well. Does it make sense for GSM-only operators to skip 3G and wait for LTE, with the potential risk of losing the competitive edge for the next several years? Should the existing 3G operator stop investing in new HSPA development? What if LTE suffers the same technological struggles as UMTS during its early stages? These are not easy questions to answer.

The time span between the first UMTS commercial deployment and the availability of a 4G commercial network will be roughly nine years. This creates big challenges for 3G operators with existing GSM legacy networks that are planning to deploy LTE, since deploying an additional network technology will represent more cost (at least initially) and more complex network operation. It is foreseen that 3G operators deploying LTE will have to operate three technologies simultaneously (GSM, UMTS and LTE), since neither UMTS nor LTE will have full handset penetration in the short to medium term.

In this section we analyze the realistic timelines for the availability of LTE technology to help assess when would be a good time to start deploying the new network. In addition, the spectrum holding situation and marketing strategies are different for different operators, and as a result the deployment and migration scenarios will vary from one service provider to another. In this section, we discuss different factors to consider when deploying a new technology, and go through some network migration scenarios that 3GPP operators will be likely to face in their 3G/4G deployment.

9.3.1 Technology Timelines

There will be two major evolution paths among different carriers: HSPA+ focused or LTE focused. Figure 9.12 shows the technology milestones for HSPA and LTE development.

When analyzing the technology timelines it becomes clear that although a lot of efforts have been put into the development of LTE/SAE by the wireless community, the real industry-wide LTE deployment will not happen until the 2011 to 2012 timeframe. There are several reasons

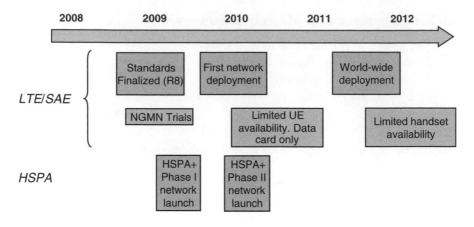

Figure 9.12 HSPA and LTE deployment time line

for that. First, the terminal device ecosystem hasn't been well established and it can be expected that this will take two to three years after the network technology is ready, based on historical precedent with other new air interfaces. Second, the ongoing UMTS deployment and network upgrades have drawn major resources from operators: for many of them, a new network was just added on top of their existing GSM network, and a lot of training and optimization work have yet to be done. Typically it will take several years for operators to get comfortable with their 3G network and gain full advantage of what the network can offer. So in reality, the operators' readiness for a new technology may be several years delayed compared to what is presented in Figure 9.12.

In general, the technology roadmap represents the industry view of the product development milestones, more from a research and development point of view and mainly driven by the infrastructure and terminal device vendors. Factors related to operators' deployment and market penetration normally are not considered. Understanding this difference will allow operators to plan capital spending more effectively and build a solid short term deployment plan which could also be beneficial for the long term network evolution strategy.

There is also the question of how to cope with the growing data demand. Based on our analysis, the safest bet for 3G operators is the HSPA+ technology, at least for the short to medium term (next five years). According to 3G Americas, in mid 2008, the number of HSPA capable device models around the world reached more than 200. With the current price of the UMTS terminals and the fast substitution rate we are experiencing, it will be possible to achieve full UMTS handset penetration in the next three to four years. So for operators who have deployed a UMTS network or have been actively building one, their main focus should be on driving new features for HSPA to leverage the maximum possible benefits offered by the 3G network. As we have demonstrated throughout this book, HSPA has great potential to deliver compelling data applications, especially with the new features which will be introduced in Rel.'7 and later product. Furthermore, the previous sections of this chapter have shown that the performance of HSPA is not far from that offered by LTE in its initial stages. Therefore the life span of HSPA will probably be longer than what appears on the technology roadmap shared by the industry.

LTE/SAE, which will be available in the 2010 time frame for commercial network deployment, will need some time to offer a competitive handset portfolio and full market penetration for operators to take full advantage of the benefits provided by the technology. Since early LTE terminal devices will be mainly data cards, the target applications will be running on computers and therefore, will not be able to offload the voice and data traffic which are carried by the mobile devices. Since the major focus has been put into data card development, early LTE chipsets will not be designed for handsets, and many important design considerations such as power consumption – which are essential for handset devices – are not taken into account. In most vendors' roadmaps, chipsets which are suitable for handset devices will not be available until 2011–2012. Taking into account the typical handset development cycle, it is very likely that there will not be a significant amount of LTE capable handsets available in the market until 2013–2014. This is closer to a real timeline that an operator can rely on to gain full benefits of the new technology when looking beyond 3G.

9.3.2 Key Factors for New Technology Overlay

Obviously, to overlay a new technology on an existing network is a complicated process which requires carefully planning on many aspects: it is not a pure technology decision. There are many factors which help determine how and when an operator should deploy the new technology, including the following: spectrum availability, cost of hardware, ecosystem, etc. Since each operator has its own unique situation, it is expected that their deployment strategy – which is driven by those factors – will be different. The list below summarizes those factors that are most critical to the operators to consider in the decision process:

- cost of new deployment;
- ecosystem (economies of scale);
- demand on data services;
- competition;
- maturity of the technology.

In the rest of this section we review each one of these aspects, and later in Section 9.3.3 we analyze some overlay scenarios with consideration of both HSPA and LTE in the picture.

9.3.2.1 Deployment Cost of the New Network

The cost associated with the new network deployment will come from five major sources: (a) the spectrum, (b) the infrastructure, (c) site development, (d) terminal development and (e) operation cost.

The spectrum is the most valuable asset for operators. In many cases, the operators cannot re-farm the existing holdings and would have to purchase new spectrum which in general is very expensive. Using the recent FCC Auction 73 in the USA as an example, the total winning bid was around $20 billion.

Infrastructure includes radio access, core and ancillary costs associated with the deployment. Existing ancillaries such as antenna, tower mount amplifier (TMA) are frequency dependent. If the new technology is deployed on a new frequency band, either new equipment is needed or the existing infrastructure needs to be modified.

Generally, site development related cost for network overlay are less than that of a green field deployment since the main facilities for a cell site such as shelter and power supply are in place and there is no initial site search activity needed. However, the cost would still be substantial depending on the frequency band the operator decides to use for the new technology.

Terminal device costs are mainly from two areas: (1) handset development related, this includes new handset testing, integration of new technology into the handset etc.; (2) operator subsidies, some operators offer discount on handsets or even give out free ones to attract new or keep the existing customers. These subsidies will have to be higher in the case of LTE, at least at the beginning, to make the devices attractive to the end-user from a price perspective.

Finally, the operator should consider the extra cost derived from operating an additional technology. As mentioned earlier, LTE will likely be coexisting with GSM and UMTS technologies in the same network, and that will represent a major challenge to the operator from a practical point of view. This involves either different crews of engineers to tackle different systems, highly skilled engineers that can cope with three technologies at the same time, or ultimately a network that is not properly run or optimized. Therefore, ideally the operator should try to consolidate their technologies and phase out at least one of the technologies (either GSM or UMTS) before operating LTE.

9.3.2.2 Technology Ecosystem

The ecosystem dictates the cost and the richness of the portfolio of terminal devices that an operator can offer. The benefits of having a good ecosystem are multifold: cheaper network equipment, less development cost for terminal devices, richer handset portfolio, increased roaming revenues and quick customer migration from the existing network.

A healthy ecosystem is determined by two factors: (1) technology and (2) frequency band. UMTS has been accepted by the majority of the operators around the world because of its backward compatibility with GSM, which was the dominant 2G technology. On top of the technology, the frequency band selection for the new technology is also critical since it is directly related to the equipment volume demanded by the whole industry. This in turn drives the equipment price and vendor's support. Unfortunately, a band selection based on building a good ecosystem sometimes may lead to more deployment cost. From an operator's long term growth perspective, selecting a good ecosystem could outweigh other cost factors such as those related to the site development.

9.3.2.3 Data Traffic Growth

Data traffic growth rate is another important factor to be considered. In the planning stage of the deployment of LTE, it is important that operators have an overall marketing strategy on applications to be delivered to the customers. The projection of traffic growth will determine the areas where new technology deployment is needed. Since HSPA and LTE are especially focused on high speed data services, in the early stages of the network evolution it only makes business sense to deploy such technology in areas where data applications are in high demand.

9.3.2.4 Competition

Competitions among different operators can also drive the deployment of a new technology. With many countries' wireless market saturated or approaching saturation, the battle for revenue growth among the carriers is becoming more and more intensive. The success of one carrier in one type of application or service typically will draw others to develop a similar product to compete for customers. Some of those applications simply cannot be offered on a 2G network, such as real-time video streaming and video calls. If an operator wants to provide those applications to its customers so that it can compete with other carriers offering such services, the only option is to deploy a 3G or 4G network capable of offering high data throughputs and low latency.

9.3.2.5 Technology Roadmap

It is important that the operator's deployment timelines of the new network are in sync with the availability and maturity of the technology. Unfortunately this hasn't been the case in the past, and some operators have suffered the consequences. The delay of UMTS commercial terminal devices is a good example: operators started deploying UMTS network as early as 2001, but couldn't materialize the benefits offered by the network due to limited availability of UMTS handsets and applications. The poor penetration of the UMTS handsets in its early stages not only stranded the capital investment of those operators, but more importantly prevented them from delivering applications which could bring in new revenue stream earlier. Therefore any network deployment plan should carefully consider when the new technology will truly be ready for mass-scale penetration (considering both network and terminal devices), and in the meantime find alternatives to cover the short to medium term demand.

9.3.3 HSPA and LTE Overlay Scenarios

Network migration, as shown in Figure 9.13, doesn't have a fixed path for every service provider. There are many contributing factors in the decision making process. In most cases, this is not a decision purely based on technical aspects. Operators' business plans and marketing strategies also play important roles as has been emphasized many times in this

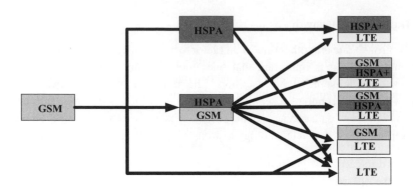

Figure 9.13   Technology migration paths for different networks

book. It is a process that requires operators to make tradeoffs between the short term plan and the long term goals.

The following section analyzes three typical migration scenarios to LTE, examining some deployment aspects as well as the pros and cons for the operators.

9.3.3.1 Migration from GSM to Partial UMTS/HSPA to LTE

This is a typical migration path for many GSM carriers around the world. Many operators normally start to deploy their 3G network in high density areas on a separate frequency band, so the 3G coverage footprint would not be as extensive as that of 2G for the first few years after the initial deployment. With the appearance of LTE, some operators may even stop expanding their 3G network and wait for LTE.

The advantages of this strategy are that the operators will be able to offer high speed data services in some areas and save capital for LTE deployment by not fully investing in the 3G network. The disadvantages are obvious because the operators may be bypassing those new feature upgrades which could substantially improve HSPA performance, they may be constrained over the data services that can be offered by the network. In addition, the potential shortage of LTE handsets could be another risk for this strategy. The realistic timing of LTE will be the key to the operators' decision.

Our recommendation for this strategy is that, with the capital saving for LTE in mind, the operator should look at the applications they are planning to offer and upgrade the HSPA network only with the new features which fit the application profiles they are offering and competitive situations they face in the market. Completely stopping upgrades to the HSPA network to save capital will run the risk of losing the competitive edge.

9.3.3.2 Migration GSM to UMTS/HSPA to HSPA+ to Partial LTE

For operators that have already deployed UMTS/HSPA in the whole network, the technology migration strategy could be different. In general, the 3G footprint is more solid for those operators since significant capital investment has been made on deploying the UMTS network. So for those operators, the focus should be on how to maximize the return on the 3G investments. HSPA will be the main packet data technology for the majority of the network. LTE could be deployed only in high density areas where HSPA+ cannot sustain the data capacity demand, or represents a higher cost per bit.

The benefit of this strategy is that the operator can offer high speed data services in the majority of its network without being constrained by the availability of the LTE capable terminal devices. In addition, as has been discussed in previous sections of this chapter, with the new standard releases, the performance of HSPA will be significantly improved. This is turn will allow the operators to extend the life span of their HSPA networks and gain maximum return of their investments. The longer life span of UMTS networks will also allow the operator to migrate the GSM users to UMTS and make the GSM spectrum available for LTE deployment. The disadvantage would be that upgrade costs for the existing HSPA network could be substantial if the operators want to stay competitive, especially if the core network needs to be re-structured for latency reduction purposes.

9.3.3.3 GSM to LTE

As we discussed, in certain parts of an operator's network, it is possible that the operator will continue to upgrade the EDGE network, skip the UMTS deployment altogether and wait for LTE. The benefits are capital savings and a simpler network structure since the operators will only need to maintain two networks. It could be a risky decision because the operator then has to use EDGE (possibly EDGE Evolution) to compete with other 3G carriers while waiting for LTE. The availability of LTE handsets will be critical for such strategy.

Applying this migration strategy, it would make it more difficult to re-farm the GSM spectrum for LTE because there will be no other technology to offload GSM voice traffic. One option is to buy additional spectrum for LTE. It is simple but expensive and also incurs extra site development related costs. The second option is to start deploying multi-mode eNode-Bs which are capable of supporting GSM and LTE simultaneously. In early deployment when LTE terminal penetration is low, the operator can start turning on LTE with the minimum bandwidth (1.4 MHz) and gradually increase it when needed. This, as we mentioned, may require migrating GSM voice traffic to LTE in certain areas to make more spectrum available for LTE, therefore VoIP support on LTE will be essential for this strategy.

9.4 Summary

Technology evolution drives the industry forward by providing new services and more spectrally efficient features. For operators, the path to network modernization is not always as straightforward as the technology roadmap indicates. It is a complicated process which leads to different network migration strategies. As we conclude this chapter, the most important message we want to deliver is: operators' business and marketing strategies are the driver for the network evolution. The gauge for success is how successfully the operators can leverage what the technology offers and move forward smoothly.

These are our main takeaways from the chapter:

- LTE technology is being almost unanimously backed by the infrastructure and operators' community as the evolution of 3G networks. This will be translated into better economies of scale. We believe that LTE will ultimately be the right evolutionary choice for cellular operators.
- One of the main advantages of LTE will be the reduced cost of the equipment, which will need limited amount of hardware upgrades and operations (simpler network) in the long run compared to 3G.
- However it will take some time for LTE to reach maturity to offer a competitive advantage over HSPA+, possibly due to limited handset offerings, and operators need to consider this risk in their strategy decision.
- HSPA+ is an excellent platform from which to offer new data services. Furthermore, the performance of HSPA+ will be competitive with the first release of LTE, further prolonging the life span of UMTS.
- To reduce the complexity of the terminals and operational costs, we recommend phasing out the GSM technology before or at the time of deployment of LTE, but keeping the underlying HSPA network until a full LTE penetration can be achieved.

References

[1] 3G UK article, '3G Market Could be Stifled by High IPR costs', 28 April 2005, http://www.3g.co.uk/PR/April2005/1384.htm.
[2] 3GPP Technical Specification 25.913, 'Requirements for Evolved UTRA (E-UTRA) and Evolved UTRAN (E-UTRAN)'.
[3] 3GPP Technical Specification 25.912, 'Feasibility study for evolved Universal Terrestrial Radio Access (UTRA) and Universal Terrestrial Radio Access Network (UTRAN)'.
[4] 3GPP Technical Requirement 23.882, '3GPP system architecture evolution (SAE): Report on technical options and conclusions'.
[5] LTE/SAE Trial Initiative, http://lstiforum.org/.
[6] 3GPP RAN1#48 Contribution R1-070674, 'LTE physical layer framework for performance verification'.
[7] 3GPP Technical Specification 25.814, 'Physical layer aspect for evolved Universal Terrestrial Radio Access (UTRA)'.
[8] 3GPP RAN1#49 Contribution R1–072580, 'LS on LTE performance verification work'.
[9] Dahlman, E., Ekstrom, H., Furuskar, A., Jading, Y., Karlsson, J., Lundevall, M., Parkvall, S., 'The 3G Long-Term Evolution–Radio Interface Concepts and Performance Evaluation,' Vehicular Technology Conference, 2006. VTC 2006-Spring. IEEE 63rd Volume 1, 2006 Page(s): 137–141.
[10] Nihtila, T., Kurjenniemi, J., Lampinen, M., and Ristaniemi, T., 'WCDMA HSDPA network performance with receive diversity and LMMSE chip equalization'. Personal Indoor and Mobile Radio Communications, 2005. PIMRC 2005. IEEE 16th International Symposium on Volume 2, Date: 11–14 Sept. 2005, Pages: 1245–1249 Vol. 2.
[11] NGMN Based Performance, Qualcomm Inc, September 2008.
[12] Holma, H., Kuusela, M., Malkamaki, E., Ranta-aho, K., 'VOIP over HSPA with 3GPP Release 7', Personal, Indoor and Mobile Radio Communications, 2006. IEEE 17th International Symposium on Sept. 2006 Page(s): 1–5.

Index